MINNESOTA ON THE MAP

MINNESOTA

DAVID A. LANEGRAN

with the assistance of Carol L. Urness

ON THE MAP

A HISTORICAL ATLAS

 MINNESOTA HISTORICAL SOCIETY PRESS

© 2008 by the Minnesota Historical Society Press. All rights reserved. No part of this book may be used or reproduced in any manner whatsoever without written permission except in the case of brief quotations embodied in critical articles and reviews. For information, write to the Minnesota Historical Society Press, 345 Kellogg Boulevard West, St. Paul, MN 55102–1906.

www.mhspress.org

The Minnesota Historical Society Press is a member of the Association of American University Presses.

Manufactured in China

10 9 8 7 6 5 4 3 2 1

∞ The paper used in this publication meets the minimum requirements of the American National Standard for Information Sciences— Permanence for Printed Library Materials, ANSI Z39.48-1984.

International Standard Book Number
ISBN 13: 978-0-87351-593-1 (cloth)
ISBN 10: 0-87351-593-5 (cloth)

LIBRARY OF CONGRESS CATALOGING-IN-PUBLICATION DATA

Lanegran, David A.
Minnesota on the map : a historical atlas / David A. Lanegran,
with the assistance of Carol Urness.
p. cm.
Includes bibliographical references and index.
ISBN 13: 978-0-87351-593-1 (cloth : alk. paper)
ISBN 10: 0-87351-593-5 (cloth : alk. paper)
1. Minnesota—Historical geography—Maps.
I. Urness, Carol Louise.
II. Title.

G1426.S1L3 2008
911'.776—dc22

2007042936

*To all those whose visions and hard work
put Minnesota on the map*

The publication of this book was supported through generous grants from the North Star Fund, the Elmer L. and Eleanor Andersen Fund, and the June D. Holmquist Publications and Research Fund, all of the Minnesota Historical Society, and from the Institute of Museum and Library Services.

ACKNOWLEDGMENTS viii

INTRODUCTION 3

1 FIRST EUROPEAN VIEWS ✺ 6

2 MAPPING AND MEASURING THE LAND ✺ 28

3 CLAIMING THE LAND: COMMERCIAL MAP PUBLISHERS ✺ 38

4 OWNING THE LAND: COUNTY ATLASES ✺ 62

5 MAPPING THE STATE: THE ANDREAS ILLUSTRATED ATLAS ✺ 72

6 CITY PLATS AND MAPS ✺ 82

7 MAPPING THE TRANSPORTATION CONNECTIONS ✺ 110

8 MAPPING THE DEVELOPING TWIN CITIES ✺ 130

9 LANDSCAPES OF RECREATION ✺ 176

10 MAPPING THE MODERN LANDSCAPE: TWO MAPS ✺ 198

BIBLIOGRAPHY AND SOURCES 202

COMPLETE LIST OF MAPS 205

INDEX OF MAPS, PEOPLE, AND PLACES 210

CREDITS 215

ACKNOWLEDGMENTS

THIS BOOK WAS FIRST SUGGESTED BY GREG BRITTON, former director of the Minnesota Historical Society Press, who wanted a publication containing significant maps to commemorate the state's sesquicentennial, in 2008. Society acquisitions librarian Patrick Coleman identified critical maps from the society's collections, and his good humor and encouragement made our work in the society's library a constant pleasure. The earliest vision of this book dates to conversations during the late 1960s with Mai Treude, former map librarian at the University of Minnesota, whose bibliography of county atlases is the basis for a chapter in this book.

Special thanks are due to several individuals and organizations for generous loans of materials to this project. Among them are Maureen McGlinn, Ramsey County Historical Society, for a St. Anthony Park map and the Souvenir of St. Paul map; Penny Bloss, of Lured to the Lake Antiques, for postcard maps; William P. Brown, Hennepin County surveyor, for the Fort Snelling survey; Betti Kamolz, Brown County recorder, for the map of New Ulm; Laura Kigin, Macalester College Geography Department, for the map of Itasca County and several postcards; Heather Lawton, Special Collections librarian at the Minneapolis Public Library, for Cleveland's plans for Minneapolis parks; Mark Leesee, of the W. A. Fisher Map Company of Virginia, Minnesota, for permission to reprint maps of the Alexandria area and a sheet from the atlas of the Boundary Waters Canoe Area; Patricia Maus, librarian at the Northeast Minnesota History Center, University of Minnesota, Duluth, for maps of Duluth and Morgan Park; Rebecca Snyder, Dakota County Historical Society, for maps of South St. Paul; Craig Solomonson, collector, who shared his extensive knowledge of Minnesota highway maps and provided maps from his personal collection; Rod Squires, University of Minnesota Geography Department, for his comments on the Public Land Survey; and Marguerite Ragnow, curator of the James Ford Bell Library at the University of Minnesota, Minneapolis.

Additional people provided support: my sisters, Prudy Cameron and Virginia Lanegran, who helped locate maps; my summer research assistant, Matt Meleca, who explored Twin Cities map collections and made useful suggestions to early drafts; Ann L. Mulfort, certified archivist, who retrieved maps from the National Archives; Charlene Roise, of Hess, Roise and Company, who provided the final reports of the historic roads project prepared for the State Historic Preservation Office; and Nancy Hof, of the Minnesota Department of Education, who provided information on nineteenth-century schoolbooks. Special thanks go to Stan Nelson, my longtime friend and classmate, who not only read the first draft of the manuscript but also facilitated the beginning of my Minnesota map collection by providing me with an Andreas *Atlas*. The book has been significantly improved by the many contributions of Dr. Carol Urness, formerly curator of the James Ford Bell Library, who wrote the first chapter and made editorial contributions throughout. A special thank you to Cathy Spengler for the design of the book and its cover. The presentation of the maps is simply stunning.

DAVID A. LANEGRAN

I THANK DAVE LANEGRAN, longtime friend and colleague in matters geographical, for calling me back to the study of maps after my retirement from the James Ford Bell Library at the University of Minnesota. John "Jack" Parker, first curator of the Bell Library, edited the journals of Jonathan Carver (published by the Minnesota Historical Society in 1976), my source for commentary on Carver.

I have been blessed in my life by associating with people who care very much about words and their usage, beginning with Jeanne Sinnen, University of Minnesota Press. Dave and I are happy that Marilyn Ziebarth, editor at the Minnesota Historical Society Press, proves that the tradition of superb editing continues to the present. We had fun doing this book.

At the historical society, Eric Mortenson did an excellent job photographing most of the maps, and Helen Newlin generously assisted in the book's production. Thanks to John Jenson for his advice and assistance with the index.

CAROL URNESS

MINNESOTA ON THE MAP

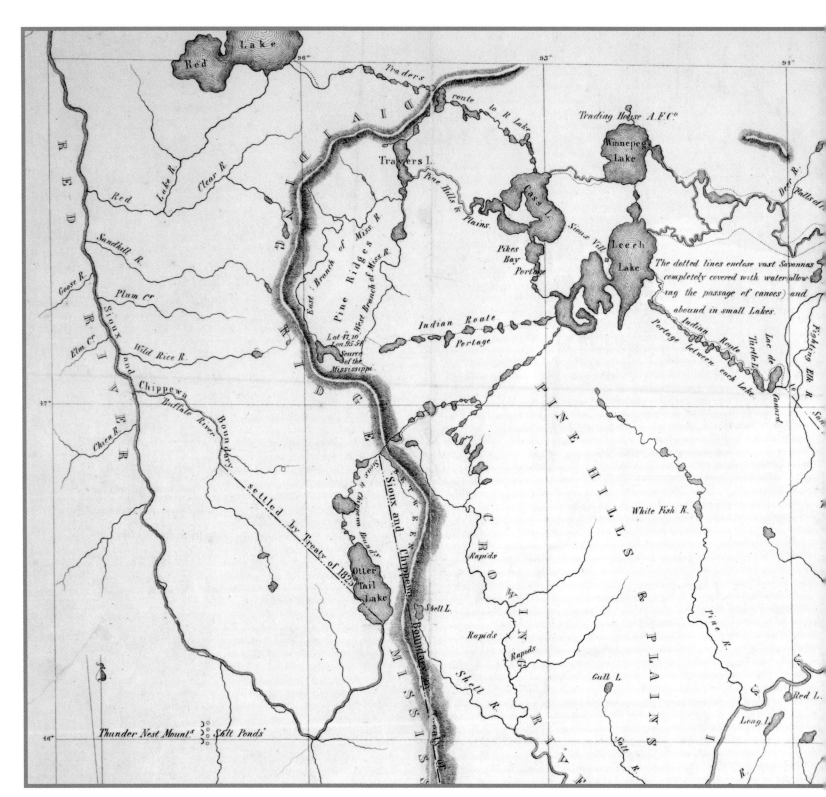

Detail, Allen, 1860 (pages 32–33)

✷ INTRODUCTION ✷

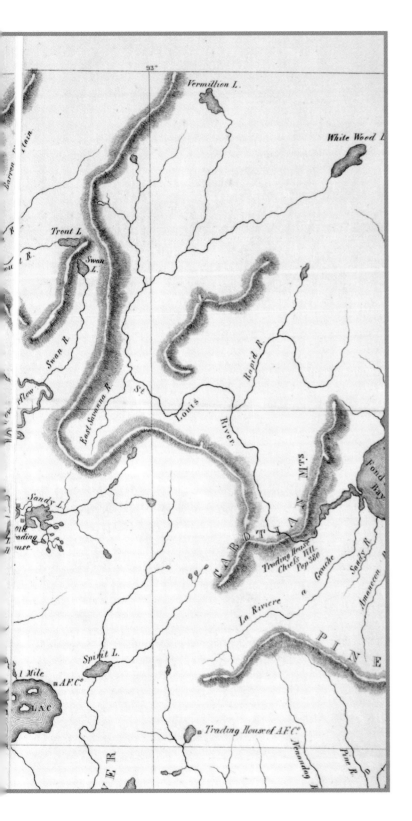

WHERE IS MINNESOTA? To answer this kind of inquiry, we look at a map. Maps answer questions about location that give us mental structure and meaning. Nearly everyone likes to look at maps, many people like to read them, and a smaller but still large number enjoy interpreting maps and uncovering their subtle messages. Maps—one of the world's earliest forms of multimedia communication, combining words, pictures, symbols, and mathematical relationships—are works of art, mathematics, and science.

Although early maps presented fanciful views of life on earth, including sea monsters, grotesque animals, and mermaids, and although Minnesota maps have shown improbable fish, talking animals, and legendary figures, most maps attempt to provide accurate locations. However distorted or embellished, maps convey a powerful sense of place. Perhaps we like them because they give us a feeling of confidence and control. At one scale, detailed maps keep us from getting lost and guide us away from danger; at another scale, they allow us to visualize distant portions of the earth's surface we cannot visit ourselves.

It is impossible to know anything about the first map, probably a sketch drawn with a stick in the dirt. Scholars debate what represents the oldest surviving map, some arguing that cave paintings, petroglyphs, and bone carvings convey spatial and location information and therefore can be called maps. Egyptian and Mesopotamian cartographers made maps of the cosmos and the earth's surface before 1000 BC on stone, baked clay tablets, pottery, and papyrus. Early urban dwellers around the world probably used maps, even though no examples survive today. Anthropologists and cultural historians have found mapping traditions in preliterate cultures.

Production of any early printed map required the contributions of skilled and knowledgeable people. Surveyors and other scientists measured and abstracted into quantitative terms knowledge about the earth's surface. Draftsmen and cartographers transformed the data into lines and symbols, and others engraved or otherwise transferred their drawings onto a printing plate. Printers produced the finished maps, which were sold as individual copies, inserted in books, or bound together with other maps into an atlas. Some individual or company marketed the finished product and financed the entire process. As we read or interpret the maps in this book, it is important to envision them within their cultural contexts and consider how the images reflect both the technology, design, and production processes, as well as the values and issues of their times.

Cultural context influenced how people mapped Minnesota. For people leaving home or for new Minnesotans, maps might have served as a first introduction to the state. Longer-time residents carried maps of the state in their heads, created by looking at maps and through personal on-the-ground experience. Even people who knew nothing about the literature and history of the state used maps in one form or another to help understand where they were.

From the wealth of cartographic representations of Minnesota from the past five centuries, we have selected a small sample that illustrates major themes. We examine the manner in which Minnesota came to be included in the expanding world map created by Europeans between 1500 and the American Revolution. Positioning Minnesota-to-be among the continents, countries, and empires of that era, maps helped legitimize the eventual conquest and occupation of lands already occupied by Native Americans. Before the land could become a commodity for sale and ownership, another generation of maps recorded its measurements and the spatial limits of political power.

Minnesota was formally organized during a transition period in the history of cartography, when new printing tech-

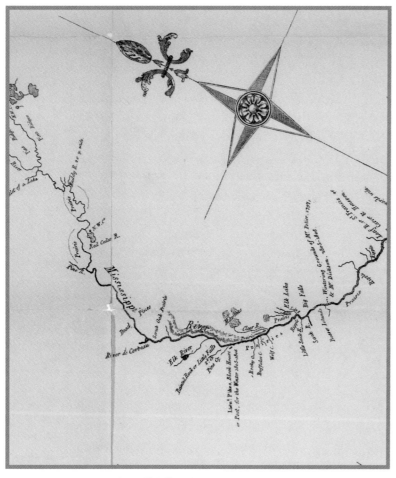

Detail, Pike, 1810 (pages 30–31)

Detail, Andreas, 1874 (pages 86–87)

nologies and marketing techniques made mass production and distribution of maps possible. Accordingly, new types of maps went to a public eager to know the location of everything from features on the landscape such as lakes and forests and political units such as counties and townships to—and perhaps most important of all—the boundaries of their own properties. Maps became legal documents, reference works, and promotional documents.

New mapping technologies kept up with the unfolding drama of the rapidly domesticating state. Following maps oriented toward agriculture, lumbering, industry, and commerce came maps that visualized landscapes of recreation and leisure. Recently, computer mapping technology has expanded the analytical power of maps enormously.

No matter how they were made or their original purpose, maps are fascinating works of science and art, economics and political power. They reflect our sense of where we stand in the cosmos and our attitudes toward other people, helping us understand our world.

The following map-filled pages show what happened in the past and what was planned for the future. Many are treasure maps of a sort, guides to mythic or dream landscapes where happiness might be found at the end of the trail. Enjoy these interesting journeys.

Detail, Twin City Rapid Transit, 1909 (pages 148–49)

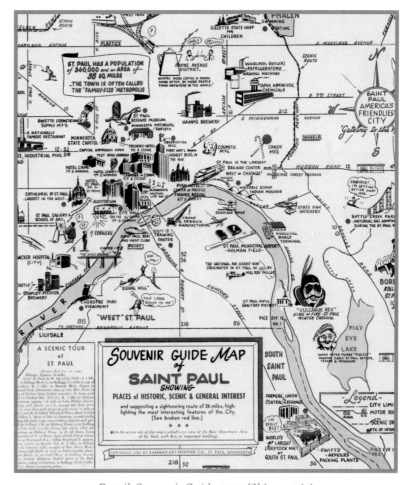

Detail, Souvenir Guide, 1953 [?] (page 167)

1

FIRST EUROPEAN VIEWS

Detail, Champlain, 1632 (page 11)

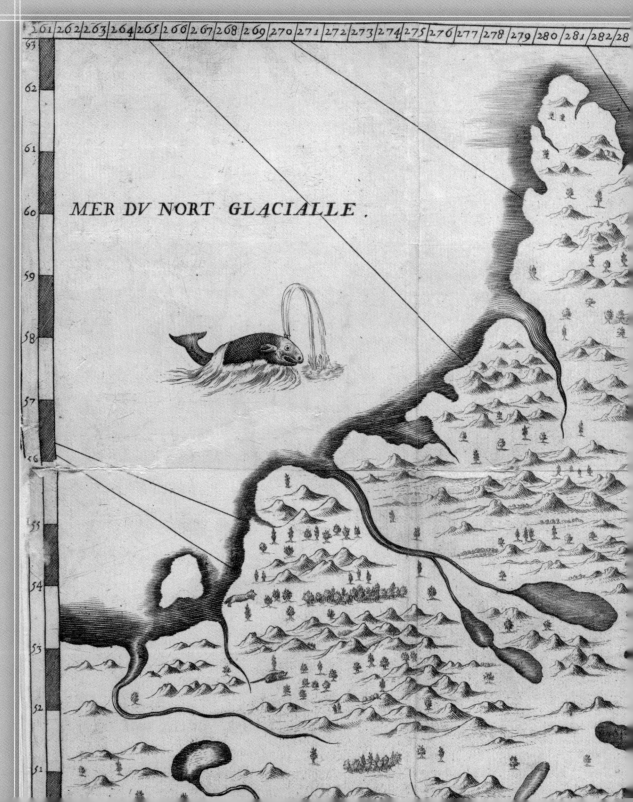

As Europe emerged from the introspective grip of the Middle Ages, the shape and interior lands of the distant New World became increasingly intriguing. Dreams of a direct route to Asia, a land fabled to hold vast riches and resources, possessed would-be explorers and their newly prospering financial backers. It is not surprising, then, that between the early sixteenth century and the late eighteenth century, the outlines and interior of the New World's continents came to be tentatively mapped and known to outsiders. By the end of this era of exploration and colonization by Europeans, a place-name resembling the word "Minnesota" appears on a map for the first time.

To prepare the earliest maps of the little-known interior region, European cartographers, most of whom never left the comforts of home, depended upon reports and sketches from European explorers, missionaries, and other travelers. These adventurous souls sketched and took notes on what they saw on their journeys; they also recorded oral information offered by native inhabitants and details from maps drawn on bark or hides.

Many early European maps have survived. With the invention of the printing press in the mid-fifteenth century, multiple copies of maps could be produced by engraving lines on a woodblock or sheet of copper and running the plate through a printing press. Printing was done on paper made from cloth and rags, and the paper could last indefinitely if well protected from fire and water. From a single woodblock or copperplate, printers made multiple copies that differed only slightly because of inking.

The first full map reproduced in this chapter, by Martin Waldseemüller, was intended to be made into a globe. It dates from 1507 and represents the earliest printed image of the entire 360-degree circumference of the earth. The globe reminds us that educated Europeans living *before* the voyages of Christopher Columbus and others knew that the earth was round. They did not know, however, how large it was. The biggest surprise for many viewers of this globe would have been the presence of a landmass clearly separating Europe from Asia, not the presence of the new name "America" that the cartographer assigned it. Europeans had hoped to reach the East by sailing directly west, but this globe revealed a sizeable land barrier between Europe and Asia.

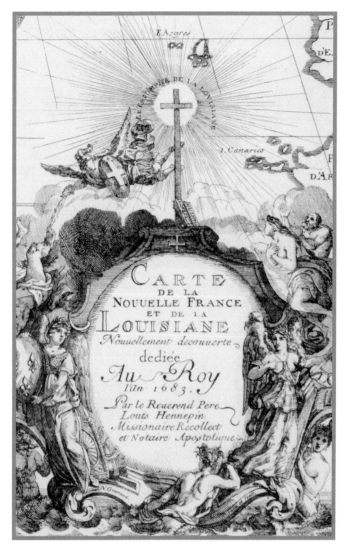

Detail, Hennepin, 1683 (pages 14–15)

This chapter's concluding map, by American-born Captain Jonathan Carver, dates to 1779, almost 275 years after the Waldseemüller globe. Like many before him, Carver sought the illusive passage to the Far East, but he was also charged with inventorying the resources of this land that the British had recently obtained from the French and that would soon be open to settlement.

Ultimately, maps—like spear points or pots—are historical and cultural artifacts. By studying these rare and early visions of the future Minnesota we learn about the mapmakers, popular beliefs, and intellectual context of the times. This makes for an exciting mental voyage.

Martin Waldseemüller. The globe gores.

ST. DIÉ, FRANCE, 1507

7.5 X 14 INCHES

JAMES FORD BELL LIBRARY,
UNIVERSITY OF MINNESOTA

Martin Waldseemüller, a noted German cosmographer, or scholar of astronomy and geography, published two maps in 1507 that used the name "America." Which came first is unknown, but they literally put "America" on the map for the first time.

One Waldseemüller map was very large, consisting of twelve sections printed with woodblocks on 13-by-17-inch sheets of paper. When assembled into three rows of four printed sheets, it is nearly 8 feet wide by more than 4 feet high. At least 1,000 copies of this world map were printed, most of which were pasted onto walls of grand houses. Today, a single copy survives, a treasure of the Library of Congress (viewable on its Web site).

Waldseemüller's other map was designed to be made into a globe to accompany a textbook for university students titled *Cosmographiae Introductio* (Introduction to Cosmography). When the map was cut out from the sheet and pasted onto a wooden ball, the globe measured only 4.5 inches in diameter. Like the wall map, the globe was printed in at least 1,000 copies, probably many more. Only four are known to have survived, and only one is outside Europe. They exist because they were never cut up to make a globe, as the mapmaker intended.

Waldseemüller's maps draw in part upon information from contemporary voyages, as well as from the traditional geography of classical scholars. He has rendered Africa's shape accurately, indicating awareness of the Portuguese voyages of exploration, but Asia is jumbled. The name "India" appears in three places; the easternmost of these, "India Superior," lies north of "Oceanis Orientalis" and "Java Major." "Zipangri" (Japan) is located west of the latitude scale. A large, bear-shaped land, probably North America, appears nearby, with islands to the east. Waldseemüller clearly had little information to work with. The head of the bear may represent the Cape Cod region, and the land that protrudes in the south could be part of Florida, but this is conjecture. The continent to the south is astonishing because the form of South America is generally accurate, although no Europeans are known to have visited its west coast. On this land he boldly placed the name "America."

Waldseemüller writes that Europe and Asia had been named for women and that Africa was named after the Latin word for "sunny" and the Greek word meaning "not cold." He continues, "Inasmuch as both Europe and Asia received their names from women, I see no reason why any one should justly object to calling this part Amerige, i.e., the land of Amerigo, or America, after Amerigo, its discoverer, a man of great ability" (Waldseemüller, 70). Explorer Amerigo Vespucci (1452–1512) made four voyages to America and wrote popular books about his adventures.

Criticism came swiftly from partisans of Christopher Columbus about the use

*Facsimile globe,
Greaves and Thomas, Isle of Wight*

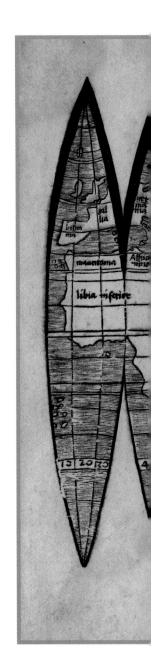

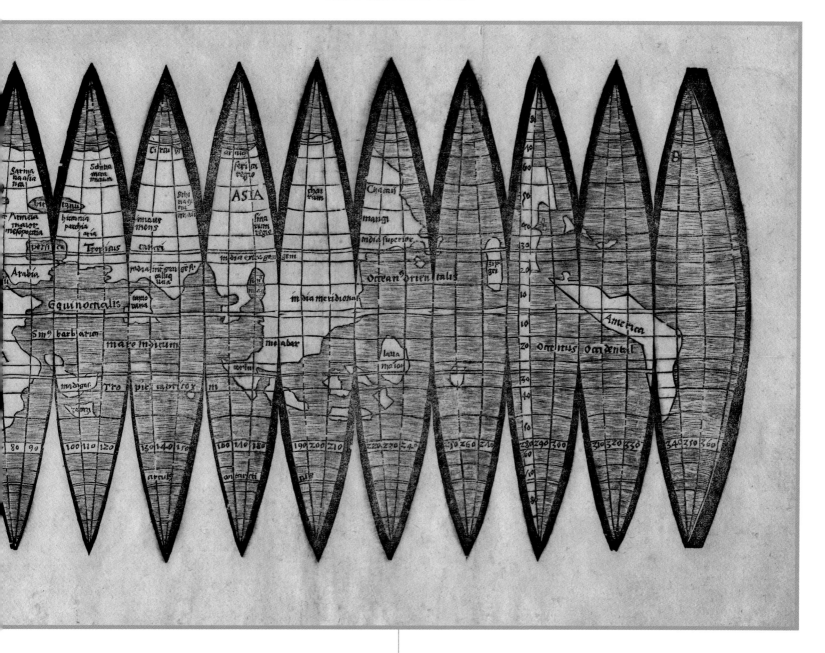

of the name "America," however, so Waldseemüller declined to use the term on later maps. Other mapmakers, however, picked it up, and it stuck, though only for the southern continent. In 1538, Gerard Mercator introduced the new name "North America" on his map of the world.

On Waldseemüller's globe, new lands—large ones—occupy the space between the continents of Europe and Asia for the first time. In addition, large oceans now separate the continents. Europeans could still hope to find a route to the East, but they would need to go around or through the highly attenuated and vaguely drawn lands—note the space between North and South America, which was formerly shown as part of Asia—that blocked their way. The search for the grail of a direct passage to Asia was still going on nearly 300 years later when the name "Minnesota" made its first appearance on a map.

Samuel de Champlain. *Carte de la Nouvelle France...,* from *Les Voyages de la Nouvelle France occidentale, dicte Canada* [Map of New France, from The Voyages of Western New France, Called Canada].

PARIS: LE-MUR, 1632

20.75 X 17 INCHES

JAMES FORD BELL LIBRARY, UNIVERSITY OF MINNESOTA

In their search for a passage to the Far East through North America, Europeans explored the continent's coasts and interior waters. Two regions offered special promise. One potential route led to Hudson Bay and the Arctic Ocean, an area explored by the English and others. The second route followed the Gulf of St. Lawrence and the Great Lakes to the interior of the continent, the area dominated by the French.

Although the French failed in their attempt under Jacques Cartier to settle Quebec between 1535 and 1543, Cartier laid claim to "New France" and learned from Iroquois inhabitants that a land farther west was named "Canada." Rumors about this new region continued to circulate in Europe, and in 1604, King Henry IV, who wanted France to have a colony in North America, sent a small expedition to establish a settlement in Acadia (Nova Scotia). He appointed Samuel de Champlain as his royal geographer, charging him with mapping Acadia. Champlain, a cartographer and artist, had already traveled to America with a fur trading ship in 1603 and had published a forty-page account of his experience.

After a terrible winter in present-day Maine, the French moved to Port Royal, where they remained until 1607, so that Champlain could map Acadia, New England, and eastern Quebec. The French moved to Quebec in 1608, with Champlain in charge. Only eight of twenty-eight men survived the harsh winter, but this rough beginning led to French success in North America. This time Champlain explored areas west of Quebec, believing, like many others, that a passage to the East could be found through the Great Lakes. Champlain learned all the geography he could from the Huron and other Native Americans, and he sent young men working in the fur trade, the *coureurs de bois,* to live with them and glean information.

Champlain also asked the Recollets, a branch of the Franciscans, to send missionaries to the region known as New France. Four of them arrived in Quebec in 1615. By the 1620s, Frenchmen had reached the shores of Lake Superior.

Champlain's first book on New France, which covered his experiences between 1604 and 1613, appeared in 1613 and contained three large folded maps and eight plates. Almost all of the geography depicted on the general map west of Montreal came from Native Americans. The map, dated 1612, names a "Gran Lac" at its western edge, which the French had not yet visited. In 1619, Champlain published a book about New France written between 1615 and 1618. His last book, including the map reproduced here, is a collected account of his expeditions.

In addition to recording Champlain's own explorations of the St. Lawrence River and Lakes Huron and Ontario, the map records information from French explorers like Étienne Brulé, one of Champlain's employees. Champlain acknowledged, "I had much conversation with them [the Huron] regarding the source of the great river and regarding their country, about which they told me many things, both of the river, falls, lakes and of the tribes living there.... They spoke to me of these things in great detail, showing me by drawings all the places they had visited" (Goetzmann and Williams, 38).

The 1632 map summarizes the knowledge Champlain and his associates had gained about New France. The latitude measurements, or distances from the equator, are especially accurate. The eastern tip of Lake Superior appears on this map, and the lake's shape is much more accurate than on the 1612 map. Champlain, who served as acting governor of New France from 1625 until his death in Quebec in 1635, had accomplished his goal of learning as much as he could about its geography.

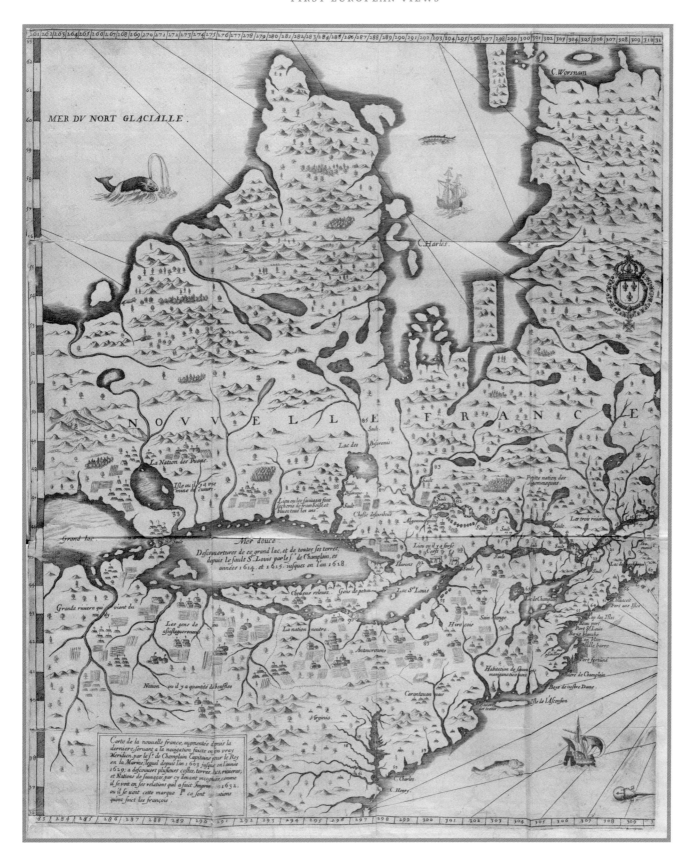

Claude Dablon and Claude Jean Allouez. *Lac Superieur et autres lieux ou sont les missions des peres de la Compagnie de Jesus . . .* , from *Relation de ce qui s'est passé de plus remarquable aux missions des peres de la Compagnie de Jesus en la Nouvelle France, les années 1670 & 1671* [Lake Superior and Other Places Where the Missions of the Fathers of the Society of Jesus Are Located, from Account of the Most Remarkable Events That Happened at the Missions of the Fathers of the Society of Jesus in New France in the Years 1670 and 1671].

PARIS: SEBASTIEN MABRE-CRAMOISY, 1672

14 X 18.5 INCHES

JAMES FORD BELL LIBRARY, UNIVERSITY OF MINNESOTA

In the years after Champlain's mapping, missionaries became the major source of information about explorations of New France. From 1632 to 1673, Jesuit missionaries living in North America wrote accounts of their experiences that were published as annual *Jesuit Relations*. Readers in Europe wanted to know about the lands and peoples of North America, and the Jesuits obliged because the sale of books supported the missions. While busy fur traders recorded little about their activities, the Jesuits wrote faithfully, and their writings became important documents for historians. Two Jesuits made this map to accompany the *Relation* of 1670–71.

Claude Dablon, who wrote this particular *Relation,* says of the map that "it was drawn by two fathers of considerable intelligence, interested in research and very accurate, who wished to set down nothing they had not seen with their own eyes. For this reason they have shown only the beginnings of Lake Huron and Lake Illinois. . . . Because they do not have first-hand knowledge of some parts of these lakes, they prefer to leave the work partly imperfect rather than to issue it with inaccuracies" (Mealing, 122).

Claude Dablon (1619–97) and Claude Jean Allouez (1622–89) explored the Great Lakes separately. Allouez, vicar general of the western missions who served the Jesuits in New France for more than twenty-five years, accompanied six Frenchmen and 400 Native Americans to Lake Superior in September 1665. He traveled along the southern shore for a month. About the lake he records: "Fish are abundant there, and of excellent quality; while the water is so clear and pure that objects at the bottom can be seen to the depth of six brasses [a measure related to arm length]" (Mealing, 95). Noting that pure copper was found on the lake bottom, Allouez wrote that twelve or fifteen distinct nations visited the coasts and islands of the lake to fish and trade. Allouez gave the lake the name "Lake Tracy," to honor Alexandre de Prouville, marquis de Tracy (1603–70), who had defeated the Iroquois. The name "Superior" on the lake relates to its position north of Lakes Illinois and Huron.

Father Dablon visited Lake Superior somewhat earlier. About maps, he notes: "By glancing, as one can, at the Map of the lakes, and of the territories on which are settled most of the tribes of these regions, one will gain more light upon all these Missions than by long descriptions that might be given of them" (Mealing, 97). The Mission de St. Esprit is located

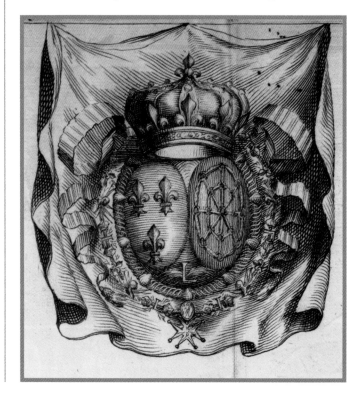

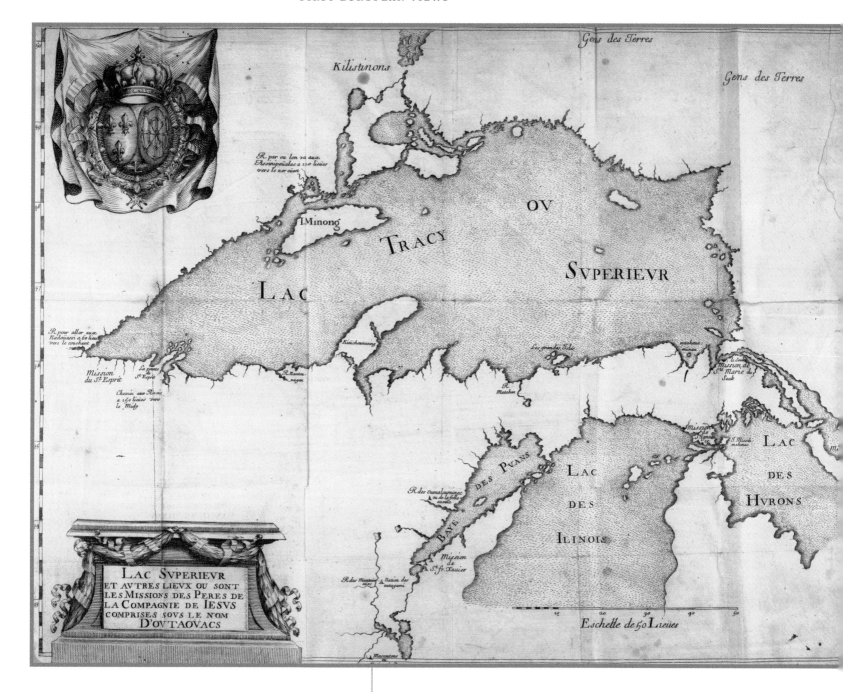

at the western end of the lake; the Mission de Ste. Marie de Sault, Mission de S. Ignace, and Mission de St. Simon are also named.

The map shows a series of bays and rivers along the north coast of the lake. According to one anonymous Jesuit, "Following a River toward the North, we arrive, after eight or ten days' journey, at Hudson Bay, in fifty-five degrees of latitude" (Mealing, 92). The route from Button Bay (near present-day Churchill, Manitoba) to Japan is easy, according to this Jesuit! "Minong" is the Ojibwe name for the place of blueberries, now Isle Royale. "Baye des Puans" is often translated as the Bay of Bad Odors. Dablon, in this *Relation*, says the origin may be the odor of the marshes around the bay, which somewhat resembles the smell of the sea. ✺

Louis Hennepin. *Carte de la Nouvelle France et de La Louisiane nouvellement decouverte au Sud Ouest de la Nouvelle France,* from *Description de la Louisiane* [Map of New France and the Newly Discovered Louisiana in the Southwest of New France, from A Description of Louisiana].

PARIS: VEUVE SEBASTIEN HURÉ, 1683

10.2 x 17.3 INCHES

MINNESOTA HISTORICAL SOCIETY

Father Louis Hennepin's 1683 map, which accompanied his book *Description de la Louisiane,* has many errors obvious to modern viewers. The east-west span of North America is exaggerated. "Mer Vermeille" (Gulf of California) extends north to nearly 40 degrees latitude. The "Lac de Conty ou Erie" suffers from a southward distention and lies at the wrong angle; the southern border of "Lac Dauphin ou Illinois" (Lake Michigan) is aborted. The "R. Colbert" (Mississippi River) ends north of the Gulf of Mexico, with the remainder of its course shown with dotted lines. The Mississippi splits in the north into two long tributaries like antlers. The Falls of St. Anthony lie on the same latitude as Superior's western end. The map reminds the viewer how difficult it was to compile the geography of the Americas.

Hennepin, born in Belgium in 1640, joined the Recollect Order at age seventeen and traveled to New France, arriving at Quebec in September 1675 and serving at Fort Frontenac. In 1678, he attached himself to an expedition initiated by the French king Louis XIV and led by Robert Cavelier, Sieur de la Salle, that traveled to the mouth of the Mississippi River. In a later edition of his published travels, Hennepin accurately showed the course of the Mississippi downriver to its mouth and incorporated La Salle's travels as though they were his own. This "borrowing" brought him some disgrace in Europe. Still, Hennepin wrote fascinating accounts of his American experiences based on firsthand knowledge of peoples and places.

Hennepin traveled with La Salle from August 1679 to February 1680. Although many Native Americans informed them that the Mississippi was not navigable to its mouth, Hennepin reported that a young Illinois warrior told them it was. He also "made us a fairly exact map of it with charcoal. He assured us he had followed its course in his pirogue and there were no falls or rapids all the way to the sea which the Indians called the great lake" (Hennepin, 81). La Salle sent Hennepin and two companions, Michel Accau (called "Ako") and Antoine Augel (nicknamed "le Picard"), north from Fort Crèvecoeur to explore the river. At one point, the three avoided capture by camping on an island, counting on their dog to bark if attackers decided to swim the river at night. On April 11, a party of Dakota (Sioux) captured the men near Lake Pepin. Hennepin named it Lac des Pleurs (Lake of Tears) because "when the Indians who had taken us wanted to kill us, some of them wept all night to make the others consent to our death" (Hen-

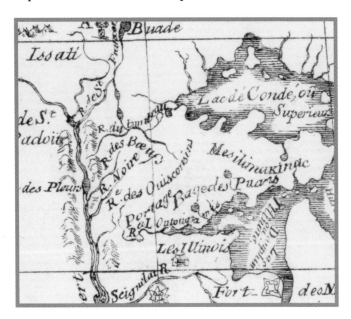

nepin, 89). The Europeans were spared, writes Hennepin, because they had a calumet, a ceremonial pipe, and other gifts to offer.

The Sioux took the prisoners north past the Falls of St. Anthony of Padua (Hennepin's name) and continued via the St. Francis (Rum) River to Lac Buade (Mille Lacs), where they spent the winter. In July 1680, while accompanying Dakota who were buffalo hunting, they met Daniel Greysolon, Sieur du Luth, and his small party, who had traveled with the Dakota to Mille Lacs. By promising to return with trade goods, Du Luth obtained the release of the three captives in late September. The Dakota chief "marked with a pencil on a paper I gave him the route we should follow for four hundred leagues of the way" (Hennepin, 125). Hennepin returned to Europe and wrote extremely popular books about his adventures before his death in about 1701.

Vincenzo Coronelli. *Partie occidentale du Canada ou de la Nouvelle France* [The Western Part of Canada or of New France].

PARIS: J. B. NOLIN, 1688

16.9 X 22.8 INCHES

MINNESOTA HISTORICAL SOCIETY

The Minorite friar and Venice-born theologian Vincenzo Coronelli (1650–1718) is remembered primarily as a maker of fine globes. In 1678, he created a large pair of beautiful manuscript globes, terrestrial and celestial, for the Duke of Parma. Because of this work, Coronelli received an invitation to France in 1681 to make an even larger and more impressive set of manuscript globes for Louis XIV. The project took much of Coronelli's time for the next two years, and his efforts resulted in globes that were approximately 13 feet in diameter, the largest known in their time.

In addition to his globes, Coronelli created maps that were engraved on copperplates for printing. In creating the map reproduced here, Coronelli had information from the French fur traders and missionaries in North America, particularly new reports on "La Louisiane" (Louisiana). The map reveals something of the international nature of mapmaking of the time. Although Coronelli left Paris in 1683, he is the primary maker of this map published in 1688. The scale of distances is indicated in Italian, French, Spanish, English, and "Lieues d'une Heure de Chemin" [leagues of one hour of the route]. The cartouche of the map notes that "Sr. Tillemon" had updated it and that the king had granted his permission for its printing.

This map features the most accurate portrayal of the location and size of the Great Lakes of any known seventeenth-century map. Lake Superior has three names: Lac de Tracy or Superieur and Lac de Conde. The people living in the present area of Duluth are the Nadouessi. Coronelli explains that this is a great nation composed of many peoples: "Issati, Tinthonha, Oudebathon, et Chongaskethon." The Algonquin Indians, when they lived near the Sioux (Dakota) in the eastern part of North America, gave the name "Nadouessiux," meaning "little snakes," to their enemies. The French shortened the name to "Sioux." By the mid-seventeenth century, these people had moved under pressure to the areas of the northern Mississippi River and Leech Lake.

The latitude of Lac Nadouessans, with four smaller lakes attached to it on the north like points on a crown, suggests that it is a representation of Lake Winnipeg, though possibly it refers to Lake of the Woods. The representation of the Mississippi River is noteworthy, ending abruptly in the northwest in an unknown place. Two smaller rivers run into it from the northeast, one unnamed and the other "La Riv. de St. Francois ou Issati, et des Nadouessions." The Prophett River, with two large lakes, connects this river to the Magdelaine River, which joins the Mississippi just south of the Falls of St. Anthony. The Mississippi River then runs straight as a pipe to the south to the approximate location of today's St. Louis, Missouri (as on the map by Hennepin above). Mountains, or at least high hills, are located along the Mississippi's west bank. To the east of the river the land is flat.

The map includes curious illustrations and several extensive commentaries about the people living along the river, making this a map with ethnographic interest. It also displays fine engraving, hand coloring that was added after printing, and unusual details. ✺

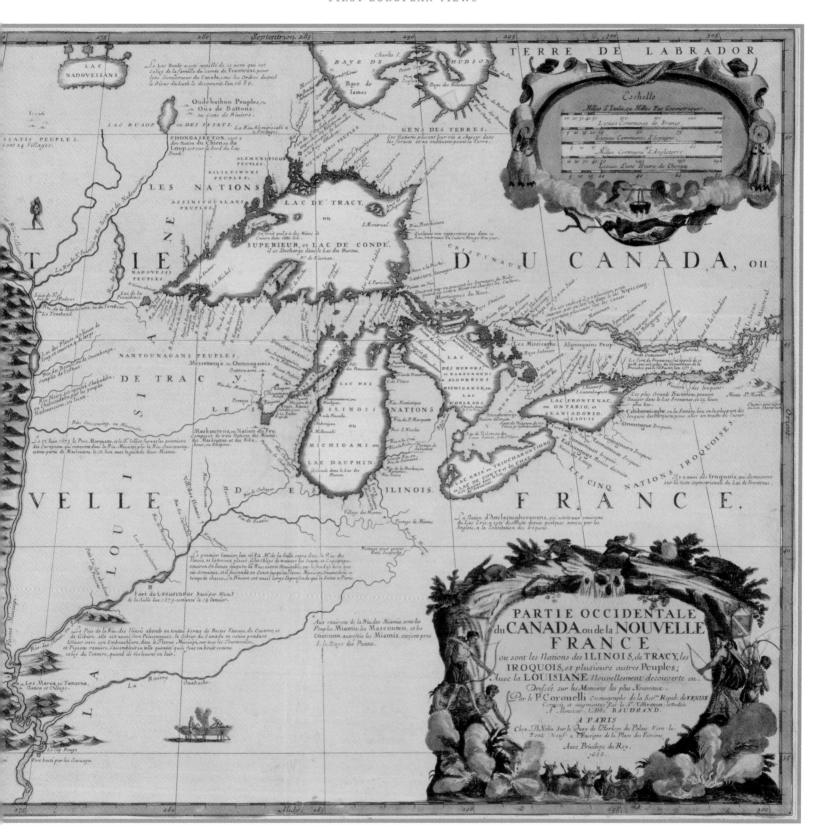

Henri Abraham Chatelain. *Carte particuliere du fleuve Saint Louis,* from *Atlas historique* [A Distinctive Map of the St. Louis River, from Historical Atlas], *vol. 6.*

AMSTERDAM: CHATELAIN, 1719[?]

16 X 19.5 INCHES

MINNESOTA HISTORICAL SOCIETY

The Chatelain firm published books and maps from 1700 until 1770, and the map reproduced here is from one of Chatelain's most ambitious efforts: a seven-volume encyclopedia published between 1705 and 1739. Nicholas Gueudeville wrote the text, which has a strong emphasis on geography. Many fine maps by Henri Chatelain accompany the text.

Born in Paris in 1684, Chatelain became a Protestant minister. He lived in London from 1710 to 1721, in The Hague from 1721 to 1728, and in Amsterdam from 1728 until his death in 1743. During his career, he produced many maps that were engraved on copperplates for printing, including some outstanding ones of North America.

The map reproduced here features the waterways of North America. In fact, a viewer's first impression is that this is a map of waterways and that the land is not important. Lake Superior is turned slightly clockwise and rounded so that its physical shape is a much less accurate presentation than on many earlier maps. This reminds us that maps do not inevitably improve in chronological sequence and that later maps are not always better than older ones. Three connections link (or nearly so) Lake Superior at the North Shore with Hudson Bay. On the map, the Mississippi River comes from the northwest at the same latitude as the northwestern corner of Lake Superior; a tributary of it joins the Mississippi farther south. This "Riv. aux Boeufs" (literally, River of Oxen, but referring to the native buffalo) does not quite reach far enough to join the "R. du Tombeau" (River of the Tomb) that flows into Lake Superior from the west. Farther south, the "R. de Ouisconsink" (Wisconsin River) flows into the Mississippi from the east.

The most special feature of this map, however, is the frame that surrounds it, which is richly filled with information. The left-hand column lists the peoples living along the St. Lawrence, around the Great Lakes, and on Hudson Bay. The text at the upper right of the map records people living in Labrador. The columns of text in the area below the map list animals, insects, birds, and fish of the St. Lawrence River and of the lakes and rivers of the interior. Also listed are shells and northern trees and fruits. Trees and fruits in the middle of Canada are added to the bottom of the right-hand column of text. The column about birds notes details of several of the continent's birds of prey and nightingales that are unknown in Europe.

The right-hand column lists goods that are traded to the Americans by Europeans. Most are weapons—guns and gunpowder, large and small hatchets, and swords and sabers, for example. Needles and thread of various sizes and colors are in

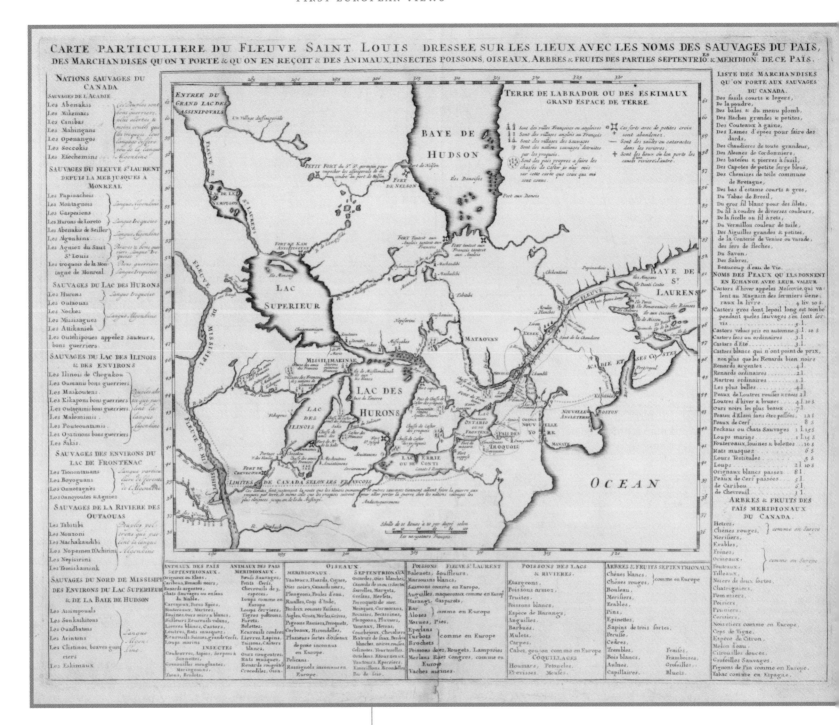

demand; soap and some kinds of cloth are also indicated, along with "tabac de Bresil" (tobacco) and "Beaucoup d'eau de Vie" (brandy). The list gives the names and prices of animal skins. Six different grades and prices are given for beaver; foxes have two prices (silver and "ordinary"). The most beautiful black bearskins fetch a high price. Prices are noted for wolves, moose, deer, caribou, weasel, mink, marten, and seals. Muskrat skins bring the lowest amount in trade. This is a fine example of a map accompanied by text that would have been highly interesting and useful for the user.

Johann Baptist Homann. *Amplissimae regionis Mississipi seu provinciae Ludovicianai* [Large Map of the Mississippi Region or Louisiana Province].

NUREMBERG, GERMANY: HOMANN, C. 1720

20.5 X 24 INCHES

MINNESOTA HISTORICAL SOCIETY

Born in 1664, Johann Baptist Homann spent his entire life in Nuremberg. He studied first at a Jesuit school and may have become a Dominican monk. The records of the city indicate that he became a notary in 1687, but his life changed completely when he became employed as an engraver.

Homann established his own mapmaking firm in 1702 and began publishing maps and globes. His first great atlas appeared in 1716; his second appeared in 1719. King Karl VI appointed him imperial geographer in 1715, the same year that he became a member of the German Academy of Sciences. When Homann began publishing, the Dutch and French dominated the atlas- and map-publishing field. By producing elaborate maps at lower prices than were charged by the established publishers, Homann, and Germany, became a strong competitor in the field of map publishing. He created maps to order based on information made available to him. Peter the Great of Russia commissioned him to make several maps of Russia according to Russian surveys and explorations.

Part of the map reproduced here, as noted in the cartouche, is based on Father Hennepin's mapping work in North America. The Great Lakes are well portrayed and the accuracy of the course of the Mississippi River is much improved over

that in many earlier maps. Clearly, Homann used the 1718 map of Louisiana by Guillaume de l'Isle in making this map also. The map's date, 1730, is obviously wrong since Homann died in 1724. His son, Christoph Homann, continued the firm until his death in 1730. After this date, the firm published maps under the name "Homann heirs." A Homann atlas appeared in 1730; the maps in it were largely reprints of earlier maps, including this one, which should probably be dated about 1720. The map was printed from a copperplate engraving, with the color added after the map was printed and dried.

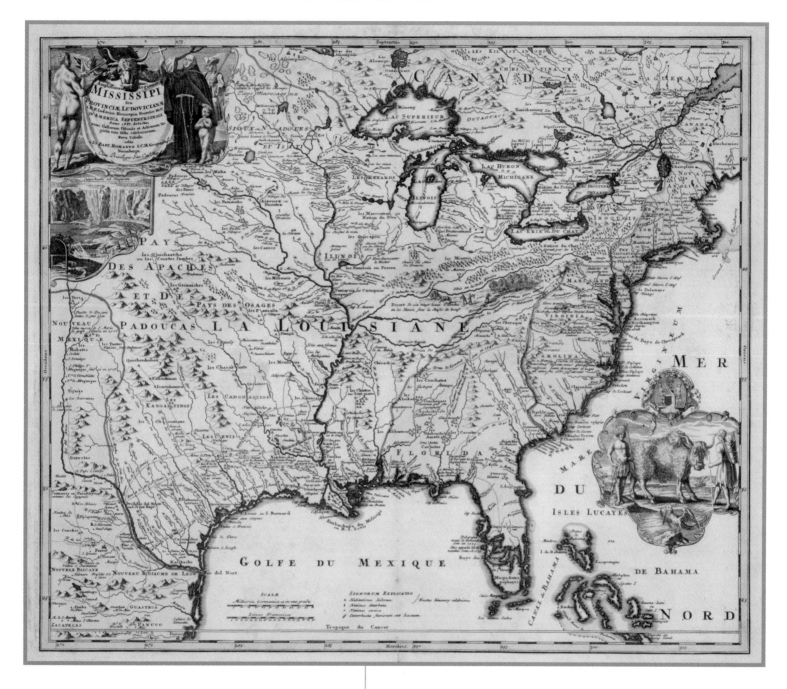

The elaborate cartouche features a missionary, presumably Father Hennepin, with his hand on the head of a child. At the center, a buffalo's head rests on the title of the map. Niagara Falls is shown below, with a bevy of beavers at work on their dams. Another feature of special interest is the "sources of the Mississippi," located near the left arm of the soldier behind the missionary. The lower-right engraving features a Native American family, with the child carried on the mother's back. In the small circle above them, an Indian couple pours something—perhaps gold—from a horn of plenty. A buffalo in profile is featured in the center of this engraving. Above the buffalo, an opossum hangs by its tail; below, a pelican raises its wings.

With these charming images, Homann captured some of the ideas that Europeans associated with North America in his time and later.

A New Map of North America from the Latest Discoveries, from *London Magazine* 32 (February): 64.

LONDON, 1763

10.6 X 15 INCHES

MINNESOTA HISTORICAL SOCIETY

In the eighteenth century, periodicals offered readers coverage of current events. One of the most successful journals of this kind in English, *London Magazine,* appeared monthly from 1732 to 1783. Its subtitle, "Gentleman's Monthly Intelligencer," indicates its intention to be a focus for discussion and a source of information on events in various parts of the world. Undoubtedly, visitors to London coffeehouses read it carefully.

In the mid-eighteenth century, wars in Europe and America were important ongoing topics for readers. For many years, French and British conflicts in Europe had spilled over to North America, and the French and Indian Wars raged from 1689 to 1763. Changes in the European empires in North America had been taking place over these years. By the Treaty of Utrecht in 1713, the British gained Newfoundland and Hudson Bay from the French (a boundary noted on the map). The map also includes the "North Bounds of New England by Charter 1620" and the boundary of the British colonies in America. When the British defeated the French in 1759 at the battle of the Plains of Abraham, the balance of power swung toward the British. Quebec surrendered soon after this French defeat; the city of Montreal was taken a year later. The long-standing French empire in North America was rapidly dissolving. When the Peace of Paris treaty ended all the fighting between France and Britain in North America in 1763, the French handed over Canada and Louisiana east of the Mississippi River to Great Britain.

The text of the *London Magazine* notes that this map "is more distinct and correct than any hitherto published." As a result of the treaty, "France gives up Nova Scotia (Acadia), Canada, and all its dependencies, Cape Breton and all islands and coasts

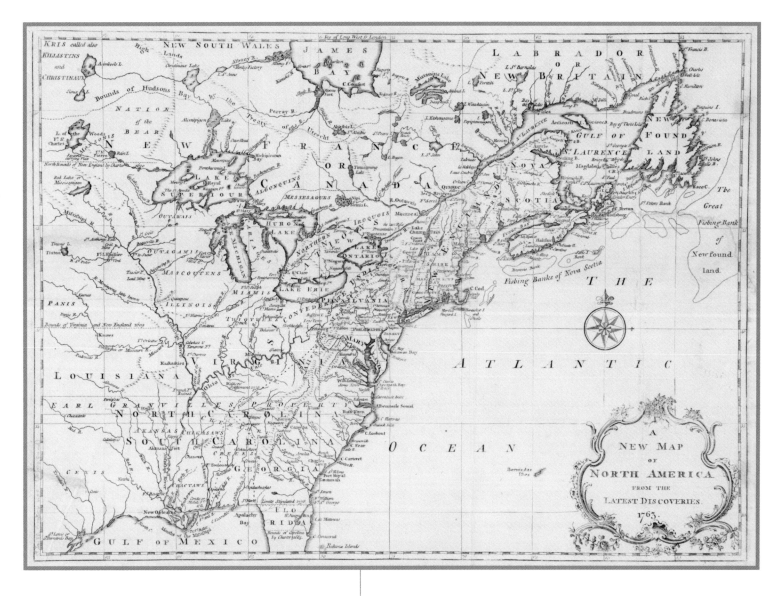

in the Gulf of St. Lawrence." The article noted that the king of Great Britain granted freedom of religion to Roman Catholics, plus the right to the French to emigrate freely for eighteen months. The French retained fishing rights in Newfoundland and the Gulf of St. Lawrence. Great Britain turned over the islands of St. Pierre and Miquelon, south of Newfoundland, to the French. The city of Mobile was ceded to the British; the French kept New Orleans. The British returned the West Indian islands of Martinique and Guadeloupe to France, which in turn gave over the islands of St. Vincent, Dominica, and Tobago to Great Britain. Spain received Cuba from Great Britain in exchange for East and West Florida. Spain also recovered Manila, and the French claims in Louisiana west of the Mississippi River were ceded to Spain. All of this had to be recorded on maps.

Because the Peace of Paris treaty was signed on February 10, 1763, and because this was the February issue of the *London Magazine,* the map may have been adapted from an already existing map (despite its title). Map features significant for the Minnesota region include Lake of the Woods, "Rain" Lake, Red Lake, and Tinton (Prairie Sioux) Lake. The St. Louis River runs into Lake Superior, which shows the named islands Minong, Royal, Maurepas, and Ponchartrain and the twelve unnamed islands constituting the Apostle Islands. Two water routes lead from Lake Superior to James Bay, indicating a continuing interest in finding a way to the Far East.

John Mitchell.
A Map of the British and French Dominions in North America.

LONDON: J. MITCHELL FOR JEFFRYS AND FADEN [1774]

8 SHEETS. 6.5 X 4.5 FEET (ASSEMBLED)

JAMES FORD BELL LIBRARY, UNIVERSITY OF MINNESOTA

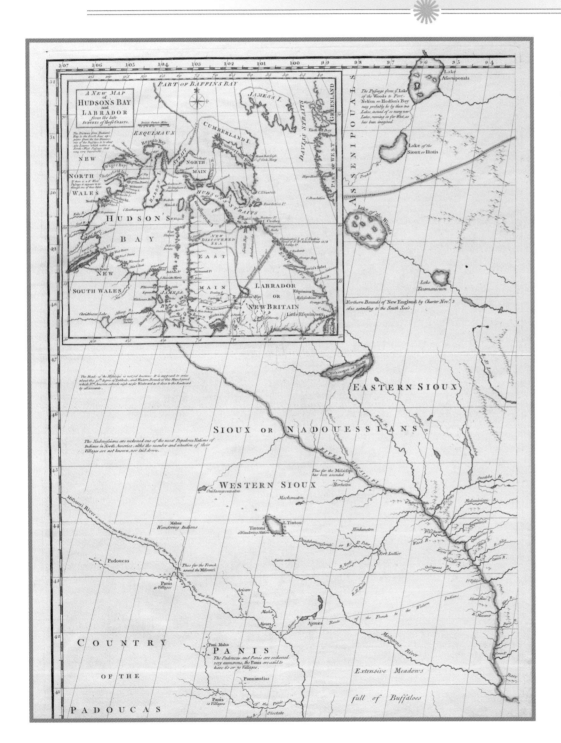

John Mitchell, born in Virginia in 1711 into a family of merchants and planters who could afford to pay for a good education, studied medicine in Edinburgh, at the time one of the best places for medical training. He returned to Virginia in 1731, where he began his medical practice. But Mitchell's favorite leisure activity was studying the flora of Virginia, and he made a name as a botanist for this work.

Mitchell argued that a series of epidemics in Virginia were due to unsanitary conditions on British troopships. When he became ill himself in 1746, he went to England for a cure. Mitchell, like other colonists, believed that the French threatened the British colonies and, in particular, that the French were encroaching on lands ceded to the British by the Treaty of Utrecht in 1713. To prove it, Mitchell began to make a map. For making his first draft, completed in 1750, he used only the published maps available to him. Learning about his efforts, the Board of Trade and Plantations hired Mitchell to make a map for it that incorporated the information that it had as the chief government agency in charge of the colonies.

The result was a large map published in 1755 under the title *A Map of the British and French Dominions in North America*. This map went through several printings. The sheets reproduced here are from the 1774 edition, which is beautifully produced on eight separate sheets printed from carefully engraved copperplates. The Mitchell map is recognized as one of the most important maps printed during the eighteenth century. The later 1775 edition was the one selected for the negotiations for the Treaty of Paris of 1783 that ended the American War of Independence. The copy of Mitchell's map with the red lines drawn on it by the negotiators to mark the boundaries of the new United States is in London at the British Library.

For the area that became Minnesota, the negotiations had an unusual result. The boundaries were tentatively set before the end of 1782, but the treaty was not signed until nearly a year later, on September 3, 1783. Much was unknown about the geography of North America, and time was needed to set the boundaries for the treaty. Mapping problems abounded in the Minnesota region, some of which are obscured by the inset map of Hudson Bay. The boundary was drawn through the Great Lakes, then to the Pigeon River and along the water route to the Lake of the Woods. From this lake's northwest corner, the border between the United States and Canada was to run west to the Mississippi, which Mitchell incorrectly showed as extending far to the north. Since the headwaters of the Mississippi, thought to lie to the west, were actually far south of the Lake of the Woods, a serious problem arose. To make matters worse, Lake of the Woods had no northwest corner. It took more than a half century of additional negotiations to settle this boundary question, and the resulting piece of U.S. territory, known as the Northwest Angle, remains today inaccessible by land from U.S. territory.

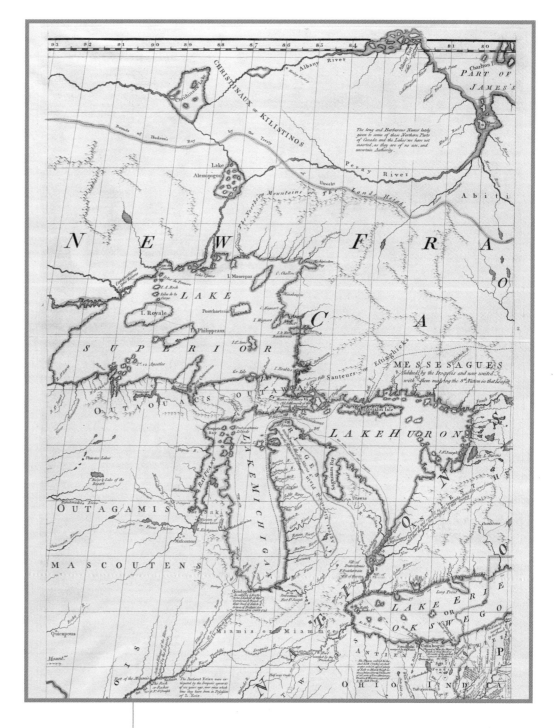

A Plan of Captain Carver's Travels in the Interior Parts of North America in 1766 and 1767, from Jonathan Carver, *Travels through the Interior Parts of North-America, in the Years 1766, 1767 and 1768*.

LONDON: FOR THE AUTHOR, 1779

12.5 x 17 INCHES

MINNESOTA HISTORICAL SOCIETY

Little is known about the education of explorer and writer Jonathan Carver. He was born in Massachusetts in 1710 and moved to Connecticut as a boy; his manuscript journals reveal that he learned to write well. After marrying in 1746, the Carvers moved to western Massachusetts, where Jonathan may have worked as a shoemaker. During the French and Indian Wars, he served in the colonial militia off and on for eight years. By the time he retired from the military as a captain, he was fifty-three years old and had seven children.

Carver bought books about surveying to help him learn the art of mapmaking. His goal was to explore the lands that the British gained from France by the 1763 Peace of Paris. At the same time, Major Robert Rogers, leader of Rogers's Rangers during the French and Indian Wars, traveled to London to petition the Board of Trade to lead an expedition to search for the Northwest Passage. Rogers noted the mapmaking skills that Carver had acquired and hired him as mapmaker for the expedition. Thus began the travels that made Jonathan Carver famous and led to the publication of his book and map.

Between 1766 and 1768, Carver traveled in the areas shown on this map, which indicates the route he took. His book is a rich commentary on his experiences. Unfortunately, an editor embellished the manuscript travel account, tainting it for many years with charges of plagiarism and deceit. Carver claimed to have made a treaty with the Dakota (Sioux) in the spring of 1767, at Carver's Cave along the Mississippi River near present-day St. Paul, by which he obtained a large tract of land bordering the Mississippi on the east. In 1780, Carver died in London, where his book was first published. After his death, his heirs made unsuccessful claims for the land supposedly granted to him.

On Carver's map, the shape of what would become the state of Minnesota almost leaps off the page. The string of lakes in the north connecting Lake Superior, Rainy Lake, and Lake of the Woods is nearly accurate. Carver notes: "This Passage leads to Lake Wenepeck and Hudson's Bay or the North West," a reference to the continuing hope of finding a route to the Pacific. The Mississippi River has tributaries coming from the east: Cold River, River St. Francis, Rum River, and Chipeway River, as well as the Falls of St. Anthony and Lake Pepin. The future southern border of the state is approximate to the boundary shown for the "Naudowessie Country." In the west, White Bear Lake, Goose Lake, Marshy Lake, and Little Lake River are named. The feature northwest of White Bear Lake is especially noteworthy: it is a small lake with a river running from it to the west, named "Heads of Origan," indicating a possible route to the Pacific Ocean.

The text on the major river on the southwestern part of the map is as follows: "River St. Pierre call'd by the Natives Wadapawmenesoter." This last name is important because it appears to be the first time a term similar to "Minnesota" is used on a map. This is near the place where Carver wintered with the Dakota (Sioux). The state's name was taken from this river, which was called St. Pierre by the French and St. Peter's by the English. On March 6, 1852, the territorial legislature requested that these names for the river be discontinued in favor of the Dakota name "Minnesota."

FIRST EUROPEAN VIEWS

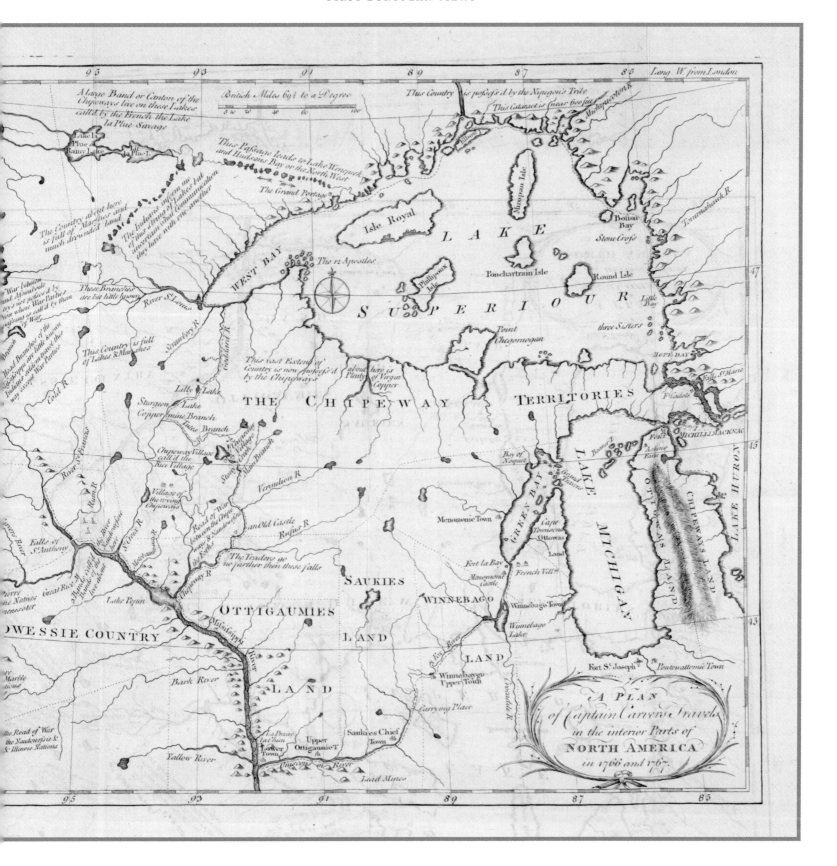

2

MAPPING AND MEASURING THE LAND

Detail, Nicollet, 1843 (page 35)

Neither the United States Congress nor the president called for mapping the extent of the new nation in the late eighteenth century, but the U.S. military urgently wanted maps of the country and specific plans for fortifications. Since the country's military academy grounded its curriculum in engineering, most military officers learned the rudiments of surveying. As a result, it was the U.S. Army that provided the first scientific mapmakers of the area known today as Minnesota.

Third president of the United States Thomas Jefferson had become uneasy about France's and England's power to block American access to the Mississippi River port of New Orleans. Accordingly, and somewhat against his principles, he arranged to purchase from France for $15 million the vast lands (recently held by Spain) containing all of present-day Arkansas, Missouri, Iowa, Oklahoma, Kansas, Nebraska, Minnesota south of the Mississippi River, much of North and South Dakota, northeastern New Mexico, northern Texas, the portions of Montana, Wyoming, and Colorado situated east of the Continental Divide, and Louisiana.

Even before completing negotiations in 1803 for the 530-million-acre Louisiana Purchase, President Jefferson had planned to send an exploring party to the Pacific Ocean. The government, however, did not have the capacity to carry out such a major project, lacking the necessary maps, instruments, trained surveyors, and cartographers. As a result, Jefferson had to assemble the expedition himself. In 1804, he commissioned Meriwether Lewis and William Clark to explore the upper reaches of the Missouri River and search for a route to the Pacific Ocean.

Four more expeditions examined different portions of the Louisiana Purchase during Jefferson's presidency. Two were thwarted by the Spanish military, but Lieutenant Zebulon Pike was able to travel thousands of miles into the new territory between 1805 and 1807. Indomitable men who were novice explorers commanded all of these early expeditions. Some possessed frontier experience, but none were trained specifically for exploration and they consequently made errors. Powerful individuals like Jefferson supported the treks, but no real governmental support existed for them.

During the two decades following the War of 1812, the War Department, particularly under Secretary John C. Calhoun (1817–25) and Secretary Lewis Cass (1831–36), sponsored several surveying expeditions into the extensive area between the Mississippi and Missouri rivers and to the Canadian border. The lack of scientifically trained officers hampered western surveys. To make matters worse, sectional jealousies and constitutional issues made members of Congress and the president reluctant to spend money on surveys.

In 1816, a few mapmaking officers—known as geographers during the Revolution and as topographical engineers during the War of 1812 and thereafter—were added to the peacetime army. These men were assigned to explore, survey, and create maps. Because there were never enough of the new engineering officers, the work frequently was conducted by other officers or contracted out to civilians. All three types of surveyors worked in the territory that became Minnesota, including two actual topographic engineers, Major Steven Long and Lieutenant John Pope.

Responding gradually to the dire need for mapping and surveying, congressional leaders passed the General Survey Act in 1824. This law, which authorized surveys for a national network of internal improvements, became the basis for the involvement by topographic engineers in the development of canals, roads, and, later, railroads. Lewis Cass, governor of Michigan Territory, promoted the exploration of the North Country between 1820 and 1832 and personally led an expedition into northern Minnesota. In spite of limited personnel and money, he made known the general contours of the region, verified copper deposits, and identified (incorrectly) the source of the Mississippi River. After geographer and ethnologist Henry Schoolcraft's success in planning and executing the basic reconnaissance of the Upper Mississippi and its major tributaries (the St. Peter's and St. Croix rivers) in 1832, Congress for the first time became a willing collaborator in furthering exploration of western lands. In 1834 and again in 1836, while Cass was secretary of war, the national legislature appropriated $5,000 for geological and mineralogical surveys of western lands. In the last great antebellum survey in Minnesota, the governor of Washington Territory, Isaac Stevens, surveyed the northern route for the transcontinental railroad under the auspices of the Office of the Pacific Railroad Explorations and Surveys, created in 1853 by Secretary of War Jefferson Davis.

Map of the Mississippi River from the Source to the Mouth of the Missouri, from Zebulon Montgomery Pike, *An Account of Expeditions to the Source of the Mississippi River.*

PHILADELPHIA: C. & A. CONRAD & CO., 1810

12.4 X 30 INCHES

MINNESOTA HISTORICAL SOCIETY

On August 9, 1805, with orders to find the source of the Mississippi River, to purchase sites from Native Americans for future military posts, and to bring back a few important native leaders to St. Louis for talks, Lieutenant Zebulon Pike started upriver with twenty men and a 70-foot keelboat. Compared to other army survey parties, Pike's group was woefully underprepared. Although he had grown up in midwestern military posts, neither he nor anyone else in the group spoke any Indian languages; no one had medical training; and their scientific equipment consisted of a watch, a thermometer, and a theodolite, a surveying instrument for measuring horizontal and vertical angles.

On September 23, 1805, Pike met with the Dakota at the confluence of the Mississippi and St. Peter's rivers. There he purchased more than 155,000 acres for a military reservation. Pike was some distance beyond the Falls of St. Anthony when the full force of winter weather hit the party. Had it not been for the help the men received from the handful of British traders in the area, they would all have perished. In spite of this assistance, Pike followed his orders and informed the traders that their posts were inside U.S. boundaries, that they should begin to pay taxes or duties to the Americans, and that they should no longer fly the British flag. Pike and his superior, General James Wilkinson, head of the U.S. Military for the Louisiana District, believed the British traders were trying to ally with the Ojibwe and Dakota, in case that border disputes resulted in another war. (These concerns came to be realized during the War of 1812, when British-led Indians fought against the Americans on the frontier.)

Pike's expedition rode the spring flood back to St. Louis on

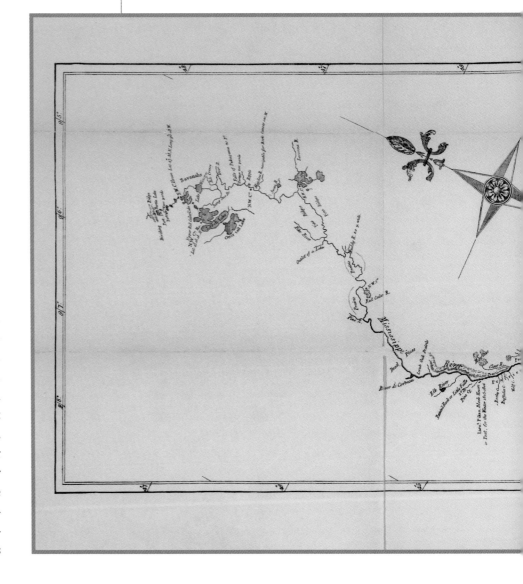

30

April 30, 1806, but Pike brought no potential native allies with him. Other than buying excellent sites for forts, he and his mission failed. Pike did not find the source of the Mississippi or sway the loyalties of the British traders, and his maps were neither well drawn nor accurate. He did, however, give his name to Pike's Peak in the Colorado Rocky Mountains, which he failed to ascend while exploring the Southwest a year later.

Lieutenant Pike failed to produce a map as a result of his travels in the future Minnesota. In 1810, however, the account of his journey up the Mississippi was combined with the journal of his more famous trip through New Spain into a massive volume published by C. & A. Conrad & Co. of Philadelphia. The publishers combined Pike's geographical observations with those of Meriwether Lewis and Mr. Thompson to produce a rather detailed map of the journey, drawn by Anthony Nau. This map was in turn redrawn by Nicholas King.

The resulting map of 1810 is the first relatively accurate map of the Mississippi as it flows through Minnesota. Little cultural information is included. The site of Pike's winter blockhouse is indicated, as are rough approximations of the land he purchased from the Dakota. The map notes the Kaposia (Little Crow) village on the Mississippi and a village very near the site of the fort. A few fur trading posts appear, as do some indications of topography, marshes, and pine forests. By the time this map was published, American Fur Company traders were already moving into Minnesota.

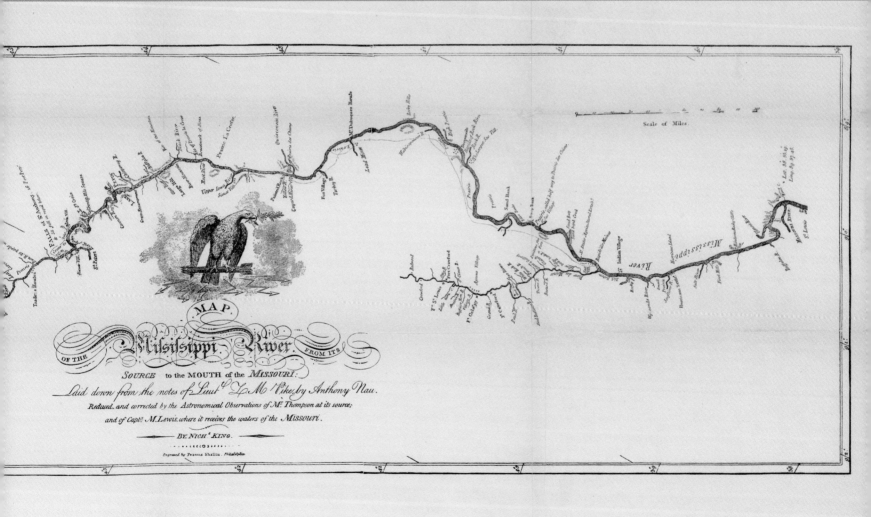

James Allen.
Map of the Route Passed Over by an Expedition into the Indian Country in 1832 to the Source of the Mississippi.
Made by Lieutenant Drayton from the original draft drawn by topographical engineer Lieutenant James Allen, from *American State Papers, Documents, Legislative and Executive, of the Congress of the United States, . . . Commencing March 15, 1832, and Ending January 5, 1836.* Vol. 5.

WASHINGTON, D.C., 1860

15 X 19 INCHES

MINNESOTA HISTORICAL SOCIETY

In 1832, ethnologist Henry Schoolcraft organized a multipurpose journey to Minnesota to follow up on the observations he had made while accompanying the 1820 expedition of Lewis Cass, governor of Michigan Territory. As Indian affairs agent to the northern tribes, Schoolcraft was concerned that the frequent Ojibwe-Dakota skirmishes would greatly hinder further settlement of the territories. Lieutenant James Allen, an infantry officer detailed to topographical duty, joined Schoolcraft to make the map and to serve as naturalist.

In addition, Schoolcraft had been ordered to conduct a thorough study of the fur trade and to vaccinate the Indians against smallpox. Schoolcraft and Allen, along with Dr. George Houghton and about thirty voyageurs and infantrymen, followed the Cass-Schoolcraft route to the head of Lake Superior. After three weeks, they reached Upper Red Cedar (Cass) Lake, where they enlisted the aid of an Ojibwe leader, Oza Windib (Yellow Head), who generously offered his assistance. Schoolcraft wrote in his journal:

> Having determined to organize a select party at this lake, to explore the source of the river, measures were immediately taken to affect it. A council of the Indians was assembled, and the object declared to them. They were requested to delineate maps of the country, and to furnish the requisite number of hunting canoes and guides. Oza Windib said, "My father, the country you are going to see, is my hunting ground. I have traveled with you many days [with the expedition from Lake Superior]. I shall go with you farther. I will myself furnish the maps you have requested, and will guide you onward. There are many rapids in the way, but the waters are favorable. I shall consult with my band about the canoes, and see who will step forward to furnish them. My own canoe shall be one of the numbers." Before night the maps were completed, and five different individuals, including Oza Windib, brought each a canoe of the proper size and laid it down. (Schoolcraft, 40)

Unfortunately Oza Windib's maps have not survived.

The Schoolcraft map shows the expedition beginning at the trading house of the American Fur Company on Big Sandy Lake. The group followed the well-established trading route through the marshes along the Mississippi, northwest to the villages on Leech Lake. The map indicates that the water was high and that the men could actually canoe through the flooded marshes. At Leech Lake, they quickly moved along another well-known route to the headwaters of the Mississippi, finding the west branch first. Many features are named, though Lake Itasca is merely called the "source of the Mississippi." The map's system of hachuring, or shading with short lines, adds a dramatic effect to the divides between the watersheds of the Red and Mississippi rivers that is not so obvious on the landscape itself. The mapmaker's hachures also exaggerate the heights of land north and west of Lake Superior,

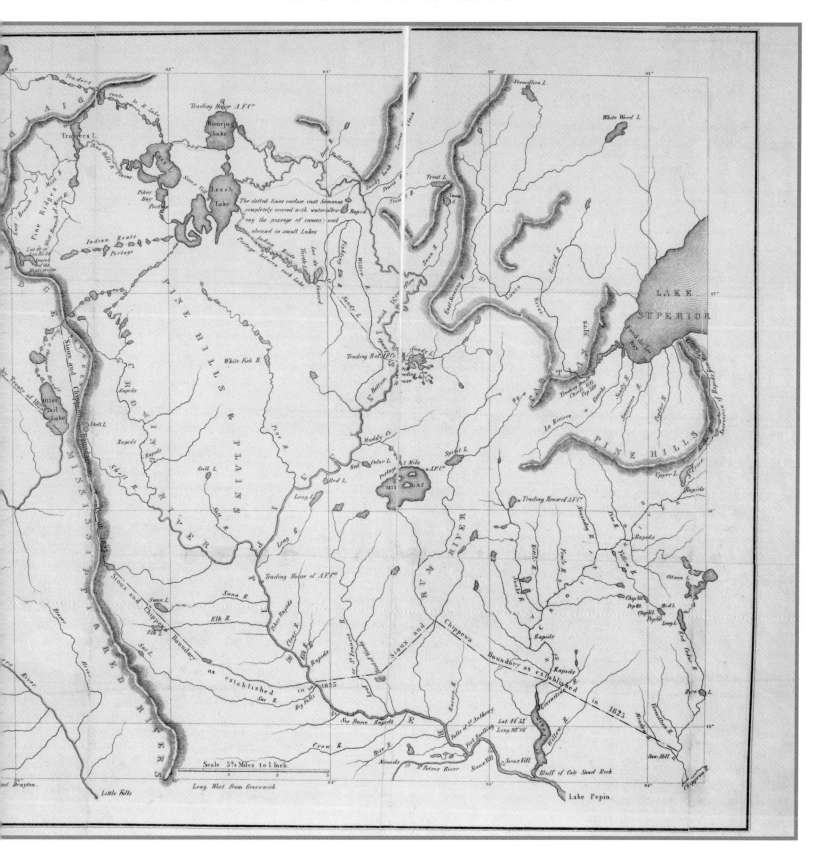

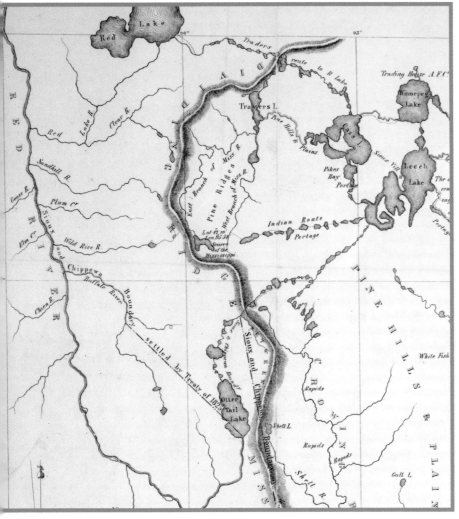

Detail, Allen, 1860 (pages 32–33)

which are named the Cabotian Mountains. This name was derived from an old name, Cabotia, which was given to all the land north of the Great Lakes by the French author Joseph Bouchette in honor of John and Sebastian Cabot, who visited North America in 1497 and 1498. The map shows a Spirit Lake that is the source of the Rum River and separate from Mille Lacs.

Joseph Nicolas Nicollet. *Hydrographical Basin of the Upper Mississippi River: From Astronomical and Barometrical Observations, Surveys, and Information, by J. N. Nicollet, in the Years 1836, 1837, 1838, 1839, and 1840; Assisted in 1838, 1839 & 1840 by Lieut. J. C. Fremont, of the Corps of Topographical Engineers, 1843.*

WASHINGTON, D.C., 1843

36.8 × 30.4 INCHES

MINNESOTA HISTORICAL SOCIETY

Joseph (Jean) Nicollet, the first thoroughly trained cartographer and geographer to work in Minnesota, was a French mathematician and astronomer. After immigrating to the United States in 1832, he decided on his own to survey and map the vast basin of the Upper Mississippi River. By astronomical and barometrical observations, he determined the geographic positions and elevations of many important points. His legacy map is recognized as one of the greatest works of American cartography in the nineteenth century.

Very much aware of his role as a surveyor, cartographer, and geographer in the creation of the new image of the Upper Mississippi, he thought carefully about the names he recorded on his map, whether indigenous, French, or American. He understood that the names preserved on it would record the elements of the natural environment and people who called it home prior to the inevitable arrival of the Europeans and Americans. Nicollet is probably the last traveler in early Minnesota to attempt to understand the region through the eyes of a Native American.

During Nicollet's first expedition in 1836, he traveled from St. Louis, where he had settled, to Fort Snelling and farther up the Mississippi to confirm Schoolcraft's earlier findings. In the course of this journey, he made nearly 2,000 astronomical and barometric observations. He corrected the earlier work of Lieutenant Allen on Schoolcraft's expedition and recorded some 300 unreported rivers, several lakes, and the routes used by the Ojibwe and fur traders as they moved through the landscape. At all times it was the Ojibwe, who knew the landscape intimately, who guided him on his journey. He wrote that he was always far to the rear on the portages from lake to lake, as he struggled to maintain his way in the difficult terrain and protect his instruments.

On August 30, 1836, leaving Schoolcraft Island in Lake Itasca and heading back downstream for Fort Snelling, Nicollet wrote:

> *We bade farewell to our pretty and romantic little island which in itself contains a complete array of all those species of trees that in this region one sometimes only encounters isolated. . . . In eighteen minutes we passed from the island to the outlet of the Mississippi, and we found the river, already a child, brisk and lively, fifteen to twenty feet wide over a depth of one foot. . . . An hour after our departure we found the Mississippi to be already thirty to forty feet wide and two to four feet deep. . . . The water is crystal clear.*

In 1838, Nicollet examined the valley of the Minnesota River and some of its tributaries. Nicollet eagerly anticipated his visit to the pipestone quarry in what is now southwestern Minnesota, which provided the Dakota with the dark red stone for the calumets that were traded across the region. His entrance into the quarry was marked by a severe thunderstorm and violent winds, as had been predicted by the Dakota.

Nicollet returned to St. Louis via the Minnesota River Valley and Fort Snelling. In 1839, he set out for the western plains aboard an American Fur Company steamboat on the Missouri River to map the remainder of the hydrological basin.

Nicollet's remarkable map was published posthumously in the year 1843. Renowned for its accuracy and elegant design, it provided a base for the surveys that followed and all the commercial atlas publishers for decades to come. Most publishers failed to acknowledge him as the source.

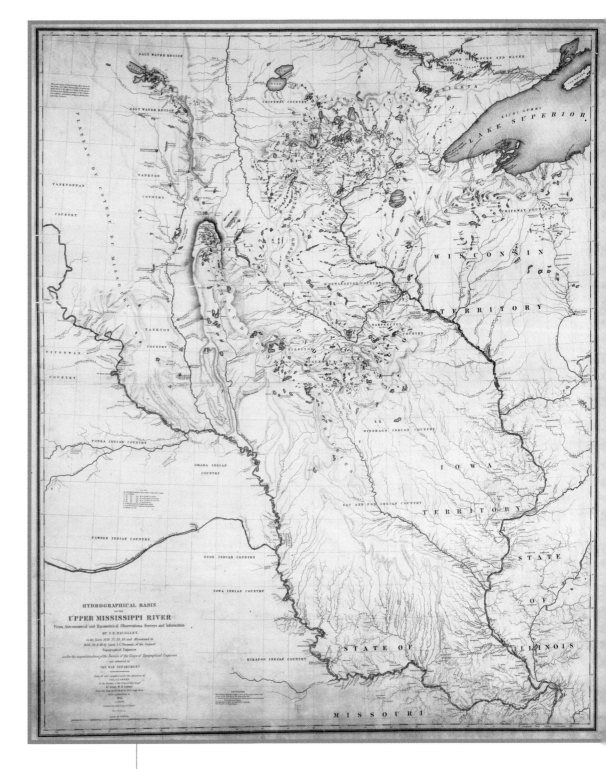

Survey of Township 103 North, Range 19 West, Freeborn County.

WASHINGTON, D.C., 1855

10 X 14 INCHES

MINNESOTA HISTORICAL SOCIETY

The Second Continental Congress's successor, the Congress of the Confederation, took two actions in the mid-1780s that profoundly impacted the development of the United States and Minnesota. Because the national government had no power to tax and had huge debts from the American War of Independence, a plan was developed that called for states to relinquish their claims to western lands, called the public domain, and for the federal government to raise money for its operations by selling them. After much debate and with bitterness, states began surrendering to the central government their claims to western lands in 1781.

Before the lands could be sold, they had to be located and measured. Under congressional mandate, a method was developed to undertake cadastral surveys and surveys of public land boundaries and to create parcels suitable for disposal by the government. The Northwest Ordinance of 1787 established a rectangular survey system of townships and ranges, which was designed to facilitate the transfer of federal lands to private citizens. The ordinance also established a process for the territories to become states and guaranteed significant rights to the residents of the territories in advance of the Bill of Rights. Following a practice in New England, the ordinance required that section 16 in each township be reserved for the benefit of public education. Thus, free public education, one of the fundamental advances introduced by the much-discussed Scottish Enlightenment, was brought to the American Midwest. One-room schoolhouses serving the population living within walking distance would eventually dot the rural landscapes of the United States.

Congress fortunately had hired an extraordinarily qualified individual to take responsibility for initiating the survey and establishing its procedures and reputation. In May 1781, Congress had appointed Thomas Hutchins as "Geographer to the United States." His first tasks were to survey the boundaries of states and the limits of federal properties. In 1787, he was identified in the legislation and given the authority to conduct the survey of the public lands.

Hutchins developed a method of dividing the land into 6-mile-by-6-mile-square townships. These townships are subdivided into thirty-six 1-mile-square sections that can be further subdivided into quarter sections, quarter-quarter sections, and, in some situations, irregular government lots. The surveys were supposed to conform to official procedures, but some errors were made, due either to honest mistakes or to fraudulent surveys. Existing surveys were considered authoritative, and any new surveys had to work from the existing surveys, in spite of errors and variations from the ideal. This sometimes resulted in sections that were far from square, or that contained well over or under 640 acres.

For the most part, surveyors in Minnesota worked on relatively flat land and had a fairly easy time doing their work. This section map, from Freeborn County in present-day southern Minnesota, shows a floating marsh, which was undoubtedly despised by the surveyors, not only because it was hard to survey but because it would have been inhabited by millions of mosquitoes. Many surveyors preferred to work in winter, when they could walk across the frozen marshes and when the biting and stinging insects were dormant.

The Public Land Survey begun in Minnesota in 1848 was finally completed in 1907, after surveyors had walked every prairie, woods, and swamp mile of the state. Today, ownership of real estate across state and local government boundaries still depends on these surveys from long ago, a surprising legacy of Thomas Hutchins and the hardy anonymous surveyors. ✺

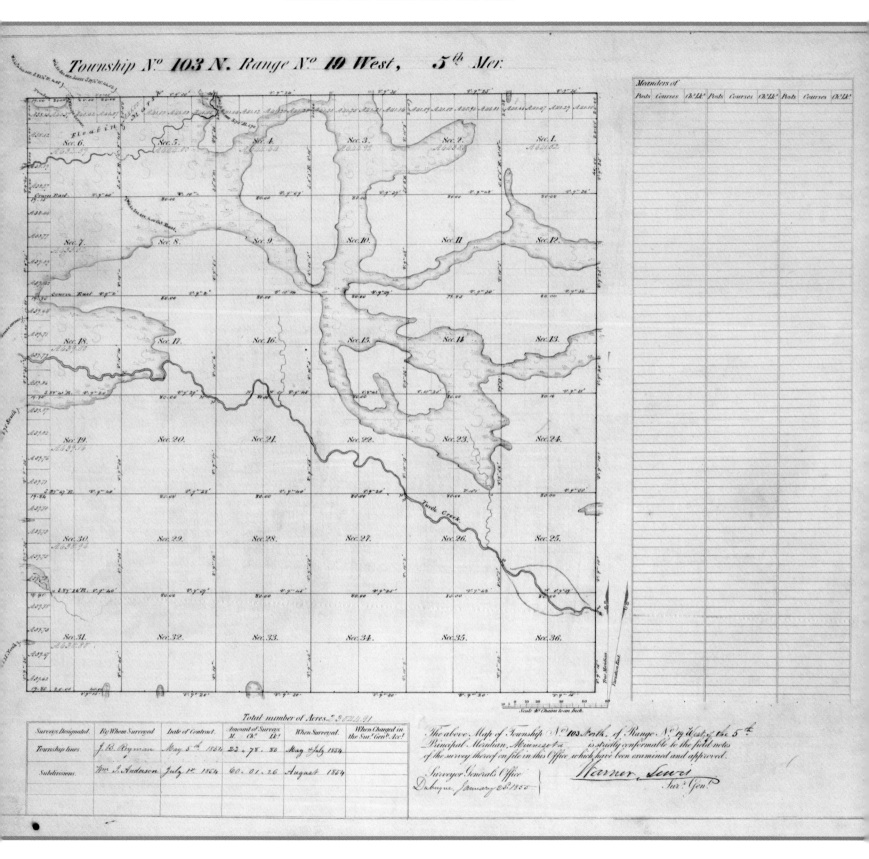

3 | CLAIMING THE LAND: COMMERCIAL MAP PUBLISHERS

Detail, Johnson, 1865 (page 59)

Between the 1830s and the 1890s, geographical knowledge about the expansive territory of the United States fairly exploded. Leaders of military expeditions, topographical engineers, and agents of the Public Land Survey filled in broad areas on maps of the Midwest, including the area that was to become Minnesota, with detailed measurements of the physical and cultural landscape. The displacement of Native Americans through forced land cessions, the creation of new territories and states, and the consequent transfers of political power required extensive and detailed mapping. The entry of new states into the Union, including Minnesota in 1858, generated the need to outline hundreds of minor civil divisions. Furthermore, new citizens needed maps to understand the powers of these emerging governments. Immigrants from Europe and newcomers from the East Coast who headed to the shifting agricultural frontier needed maps to tell them where they were and how to get to where they wanted to go next. Dynamic landscapes demand constant revision and updating of maps and atlases.

By the mid-nineteenth century, middle-class people could afford to buy maps of their home region and the world for the first time. Commercial publishers created and then met the rising new demand for geographic knowledge at all levels by making large numbers of maps and atlases available. Atlases were especially popular with the farmers and townsfolk of the Midwest. Until well into the twentieth century, people frequently learned geography within the family. Although family libraries may have been limited, by 1900 an atlas had become commonplace in middle-class homes.

The nation's growing middle class thirsted for geographical information, a need that could be met by innovative printing technologies and distribution methods. A new generation of atlas and map publishers rose to meet the challenge. The Mitchell and Colton firms dominated, but Asher & Adams, Johnson, Cowperthwait, and Anderson became significant publishers as well. These mapmakers shaped the new Americans' view of their space and of Minnesota's location within the United States.

Atlas publishers knew that their readers wanted maps of Minnesota that articulated the acquisition and taming of the landscape, so Native American claims to lands were recognized but downplayed. Land-cession treaties took on special prominence on maps of Minnesota Territory after its creation in 1849. The organization of political power became a major focus, and remarkable efforts were devoted to illustrating the establishment of counties, townships, villages, and cities. After each session of Minnesota's Territorial Legislature, maps were revised to reflect new counties, rearrangements of older counties, and the expanding Public Land Survey. Place-names were revised from Native American or French to English in new editions. Cartographers also paid careful attention to the delineation of roads and railroads, both real and proposed, and sometimes mixed them up.

Mass-produced maps that rationalized and measured the nineteenth-century world brought reassuring order to the complexities of geography. New maps appeared and reappeared, most frequently in guidebooks for emigrants, in atlases, and in schoolbooks. The guidebooks extolled the virtues of the new territories and states, and, though their maps were fairly accurate, favorable information was frequently exaggerated. Atlases presented the most complete representations of the lands, because map publishers made every effort to keep their maps up-to-date. Schoolbook maps were kept simple for young readers and usually derived from atlases, which were frequently produced by the same publisher.

Popular atlases could be global, state, or local in orientation. Publishers, of course, paid close attention to their home regions. European atlases focused on Europe, and American atlases emphasized the United States. The city of Chicago's great map and atlas publishers produced state and county atlases featuring the Midwest.

Most American maps and atlases focused on geopolitics, with boundaries of empires, colonies, and countries strongly marked. Prominent European countries received full pages, and world atlases had a northern European bias, which reflected the background and interests of the majority of American inhabitants during the nineteenth century. Maps of Africa and Asia were generalized because of the lack of specific geographic knowledge.

Publishers in the United States fanned the flames of the nationalist sentiments of their times, emphasizing the individuality of each state. Because map users were greatly interested in the country's westward expansion, states and territories west of the Mississippi were depicted as major political units—and given their own pages—even if they were only sparsely settled.

Until 1884, when England's Greenwich Meridian was formally adopted by the world's leading commercial nations in the movement to standardize global time, American mapmakers routinely drew their own prime meridian as running through Washington, D.C., as well as through Greenwich. Europeans commented negatively on this practice, but it seemed to please the American map-buying public.

John Farmer.
Improved Map of the Territories of Michigan and Ouisconsin on a Scale of 30 Geographical Miles to an Inch.

NEW YORK: J. H. COLTON & CO., 1835

24.5 X 32 INCHES

MINNESOTA HISTORICAL SOCIETY

In 1821, Michigan territorial governor Lewis Cass and the University of Michigan invited New York cartographer John Farmer to settle in Michigan. After teaching at the university for about three years, Farmer began his own successful cartographic publishing business. He eventually published twelve different maps of Michigan, Wisconsin, Lake Superior, and Detroit, and his early maps are regarded as influential in promoting extensive emigration to Michigan between 1825 and 1840. Farmer and his son Silas served as the state's principal mapmakers for the remainder of the nineteenth century. In addition, they were leaders in the county map business that developed as the settlements matured.

Farmer's map has a unique orientation that enabled him to include the large area under the nominal jurisdiction of Michigan. The caption "Michigan extends west to the Missouri River but expect a new Territory will be shortly set off as Ouisconsin" makes clear the anticipation that continued settlement would bring statehood to Michigan and separation from the western lands. Farmer's excellent map engraver was able to include a great deal of detail about the routes followed by Henry Schoolcraft, the geologist who accompanied Lewis Cass on the expedition to explore the Lake Superior region

and find the source of the Mississippi River, and Major Stephen H. Long.

An extraordinary amount of information about the Native American population appears on this map. Villages and Indian agencies are located, and Farmer even shows the sites of abandoned trading posts. In addition, he provides a sort of census of the bands, whose territory he identifies by indicating the numbers living near the trading posts. This is the only commercial map of the period that provides this amount of information about the area's early inhabitants.

Farmer only vaguely renders the subtle topography of Minnesota, using some low-quality hachuring. The river system, however, appears in great detail, with indications of width and navigability for canoes. In spite of this attention to detail, Farmer made a slip of the pen when he placed Fountain Cave on the west bank of the Mississippi River. The Dakota name for Mille Lacs, "Spirit," is mistranslated as "Rum." The map shows the triple Continental Divide near modern Hibbing, where waters drain to the south, north, and east. Farmer includes fascinating references to the forest, indicating, for example, large burned-over areas near Leech Lake, the "Forest of Burnt Dead Pines." He describes the swampy area north of Mille Lacs as a "Dreary Region."

Farmer's map perpetuates the myth of great mineral po-

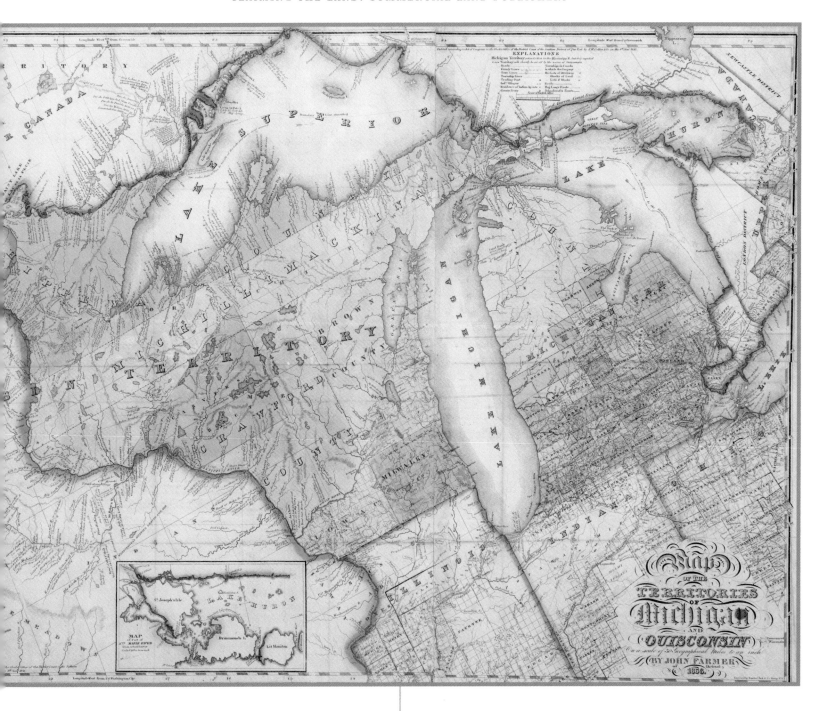

tential in the valley of the Blue Earth River by indicating a region rich with copper and lead. The counties of Michigan that would become Wisconsin extend only to the Mississippi River. Lands to the west are marked as not yet open to legal settlement, because the U.S. Army at Fort Snelling provided the only law in the region.

J. Ayers, compiler and engraver.
Map of the Settled Part of Wisconsin Territory, Compiled from the Latest Authorities, from William Rudolph Smith, *Observations on the Wisconsin Territory: Chiefly on That Part Called the Wisconsin Land District.*

PHILADELPHIA: E. L. CAREY & A. HART, 1838

22 X 17.75 INCHES

MINNESOTA HISTORICAL SOCIETY

In 1837, William R. Smith, a lawyer and Pennsylvania resident, journeyed to Wisconsin Territory with Governor Henry Dodge to observe negotiations for a treaty with the Ojibwe (Chippewa) Indians. They arrived too late, but Smith went on to tour the southern part of the territory. He kept a meticulous diary of his travels, which he used when writing the book in which this map appeared. Smith's long and productive career in Wisconsin politics began when he was appointed adjutant general of Wisconsin Territory in 1839, a position he held until 1851. He served the state in a variety of elected and appointed posts. Active in the establishment of many cultural organizations, he also published a history of Wisconsin.

Smith's 1838 book, titled *Observations on the Wisconsin Territory,* is typical of the time. It is extraordinarily detailed and full of encouragement to potential settlers. From his viewpoint, no better place to settle existed than lands recently sur-

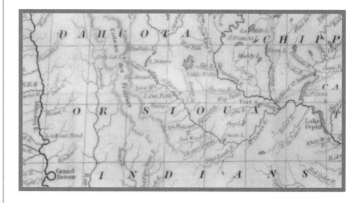

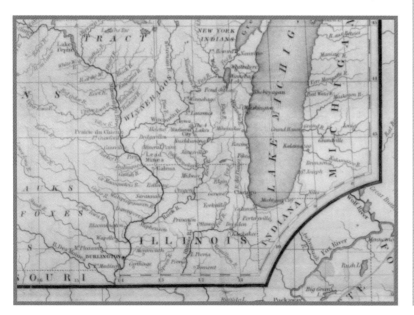

veyed and opened to settlement in this part of the territory, called "the Wisconsin Land District."

Smith's publishers probably added this map to Smith's guide for Wisconsin immigrants to help sell the book. In fact, the author suggests that readers avail themselves of maps made by competitors. Although the book discusses only the Wisconsin Land District, the map includes the rest of Wisconsin Territory as well. While it contains no new information about Minnesota, it further documents the march westward of the survey, showing settlements, roads, and towns and the removal of the Native Americans from control of their lands. The map's real focus is the land available for purchase in southern Wisconsin.

The small inset map of the entire territory distinguishes between lands where settlement was possible and off-limits lands still controlled by various Indian tribes. The cartographer clearly did not use Nicollet's map as his model, because he shows Upper and Lower Red Lake incorrectly and includes the Lake of the Woods and the mythical Carver's tract, which Carver claimed to have received in negotiations with the Dakota. Neither of these lakes, nor the Carver claim, were on Nicollet's large and detailed map.

Sidney E. Morse and Samuel Breese.
Iowa and Wisconsin, Chiefly from the Map of J. N. Nicollet, from *Morse's North American Atlas.*

NEW YORK: HARPER BROS., 1845

14.5 X 18 INCHES

MINNESOTA HISTORICAL SOCIETY

This 1845 map of Iowa and Wisconsin records changes accompanying the creation of the vast Iowa Territory in 1838. The new boundaries included a portion of the future Minnesota west of the Mississippi River. Although the map varies little from the Nicollet map, upon which it is based, it illustrates the evolving political boundaries in the Midwest and moves beyond early guidebook maps to give newly detailed information such as the location of cities and counties.

Most of the residents of Iowa Territory lived within 50 miles of the Mississippi River and near the town of Dubuque at this time. But the population was growing fast, and settlers were keen to win statehood because they would be better served by the federal legal system.

Although determining the eastern, southern, and western boundaries of the future state of Iowa was easy, the northern boundary became the subject of a great deal of argument. Many people wanted a large state and believed that inclusion of the powerful Falls of St. Anthony would greatly enhance their fledgling economy. Others were not convinced that the Dakota would surrender their rich lands west of the Mississippi in future southern Minnesota. They further thought it unwise to have such a large state occupied by a small population because, among other things, they would be saddled with higher taxes. After several organizing conventions and congressional vetoes of the residents' early plans, Iowa became a state in 1846, with its present borders.

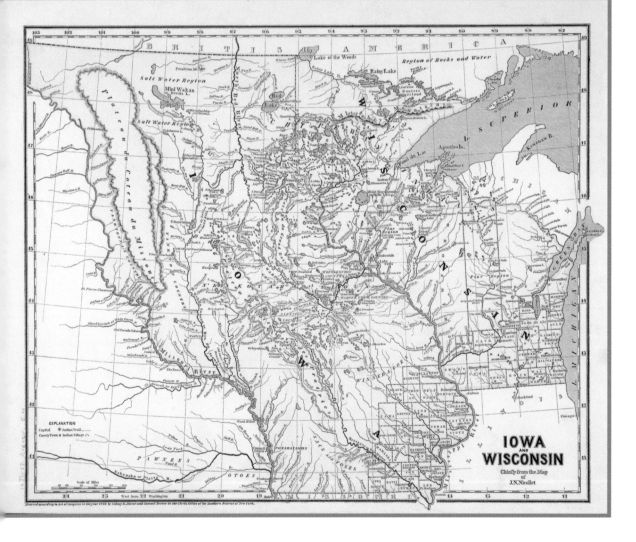

H. N. Burroughs. *United States,*
from Augustus Mitchell, *New Universal Atlas.*

PHILADELPHIA: 1848 OR 1849

17 X 14 INCHES

AUTHOR'S COLLECTION

Although dated 1846, this map must have been published during the ten months between May 29, 1848, when the state of Wisconsin was created, and March 10, 1849, when Congress created the Minnesota Territory. During these months, the area between the St. Croix and Mississippi rivers was in political limbo. It took about two years of negotiations to establish the western border of Wisconsin at the St. Croix River. About 5,000 non-Indian people inhabited Minnesota Territory at the time.

The strongest economic areas of Wisconsin were along the shores of Lake Michigan and in the lead-mining district in the southwest corner, around the old population center at Prairie du Chien. However, the rich potential of the pineries in the St. Croix watershed was well known, and Wisconsin expansionists wanted to secure the river's entire watershed and, if possible, the Falls of St. Anthony. Residents of St. Croix Valley and Mendota, however, felt terribly far away from the capital in Madison, separated by miles of undeveloped forest, lakes, and swamps. Furthermore, as Wisconsin was moving toward statehood, a bill was introduced in the House of Representatives in 1846 to create Territory of "Minasota." The House passed the bill, but the Senate did not, stating that the population of the territory was not large enough to meet the minimum requirements. In 1848, when Wisconsin became a state, Senator Stephen Douglas introduced a bill to create Territory of Minnesota, but Congress adjourned in August and did not take action until the following March. Thus resulted the extraordinary brief political snapshot depicted in this map: the state of Wisconsin doglegs to the present-day eastern border of Minnesota, and the yellow-toned jigsaw puzzle piece is unnamed.

On the opening day of the 1848 congressional session, Senator Douglas announced his intention to introduce a bill creating Territory of Minnesota, but the bill did not reach the floor until January of 1849. Although many congressmen doubted the viability of Minnesota, they passed the Douglas bill with the stipulation that the territory would be created on March 18, 1849, so that the newly elected Whig president, Zachary Taylor, could appoint the territorial officers. The new Minnesota Territory included the residues of the territories of both Wisconsin and Iowa.

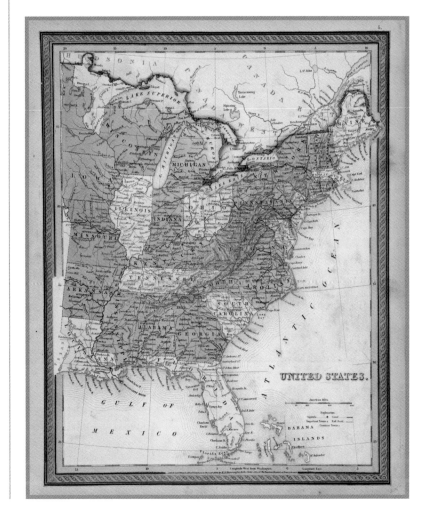

Map of the Organized Counties of Minnesota. With an inset *Map of Minnesota Territory.*

PHILADELPHIA: THOMAS COWPERTHWAIT & CO., 1850

22 X 17.75 INCHES

MINNESOTA HISTORICAL SOCIETY

Issued the year after Minnesota became a territory, this counties map, based largely on the maps of Joseph N. Nicollet and John Pope, includes extensive notes on early exploration and topography and adds new political and demographic information. On the eastern side, Minnesota's organized counties extend northward from the Mississippi River to north of Mille Lacs (Spirit Lake). St. Paul's population is 1,500; St. Anthony Township's is 900. No population is given for St. Anthony City, though the mapmaker included symbols for extant buildings.

Outside the population center, Sauk Rapids has 300 people. Fort Ripley, named Fort Gains on the map, has one company of military dragoons and one of infantry. The long and rough Red River Trail heads northwest along the east bank of the Mississippi, crossing it at Sauk Rapids and again just above the confluence with the Crow Wing or Kagiwigwan River. The map identifies Ojibwe villages and trading posts.

St. Croix Valley is much more developed, with a series of towns and sawmills ("saws"). Point Douglas has a population of 350; Willow River has 300 inhabitants and two saws. (The inset territory map shows this town in a different location north of Stillwater.) Stillwater's population is illegible, but it may read 1,000. Marine Mills has four saws and 150 inhabitants. Nearby Osceola, Wisconsin, has 100 residents and two saws. St. Croix Falls, with 500 people and eight saws, is identified as the head of steamboat navigation.

The progress of the Public Land Survey is obvious on the map, which includes not only township lines but also surveyors' comments on prairie vegetation and water features. The territory's primary resource is identified as large tracts of pine timber.

On the inset *Map of Minnesota Territory,* the formal boundary stretches westward to the border of Missouri Territory on the Missouri River, and lands of several Native American groups are identified. In the far northwest, Assiniboine country is identified, though no boundaries are shown. North of the organized counties, "Chippewa Country" appears, but no groups are identified. A Winnebago agency is sited on the Long Prairie River, and lines that appear to mark the edge of the Winnebago Reservation west of Fort Gains are present. This reflects the proposed plan to relocate the Winnebago from southern Wisconsin to a reservation created between the warring Dakota and Ojibwe, to serve as a buffer.

Elegant hachures show the Plateau du Coteau du Prairies, a plateau, or height of land, on the prairie. The map also indicates the Missabay Heights and a Coteau Grand du Bois (plateau of woods) forming the north side of the Minnesota River Valley. Apparently the mapmakers were unaware of the Sawtooth Mountains along the North Shore of Lake Superior.

These maps show a transition in the language of Minnesota's place-names from the Europeanized Dakota and Ojibwe and French names to English. The name "Mille Lacs or Spirit Lake" appears on the main map, but on the inset map of Minnesota Territory it has been simplified to Mille Lacs. Names of the Mississippi tributaries in southeastern Minnesota are intriguing. Nearer to St. Paul, they are in English (Vermillion and Cannon), but the more distant Zumbro River is still the Wazi Oju, and the Whitewater River is Miniskah. The Root River has two names, the Hokah or Root.

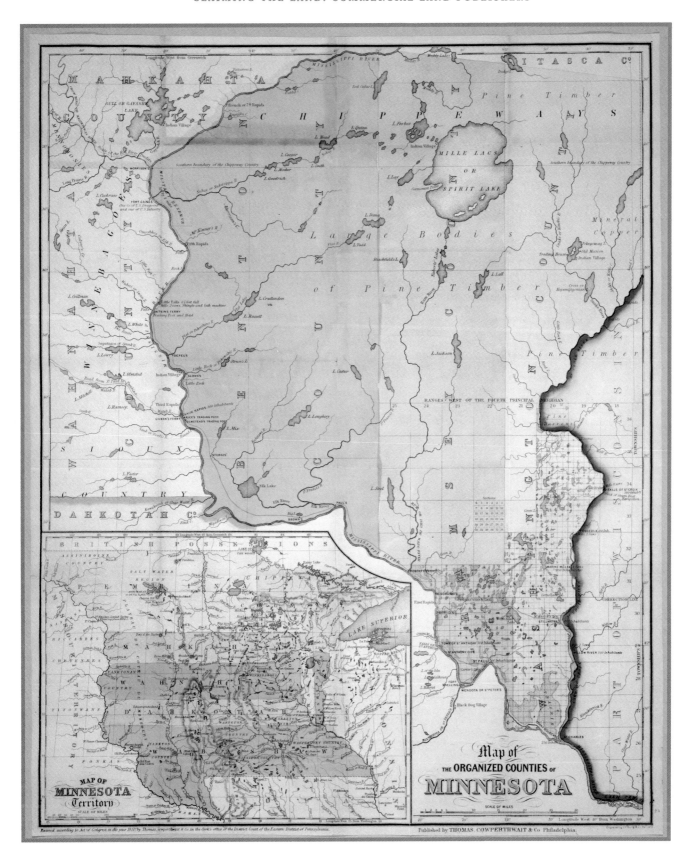

J. H. Young. *Map of Minnesota Territory*, no. 36, from *Mitchell's Universal Atlas*.

PHILADELPHIA: THOMAS COWPERTHWAIT & CO., 1852

13.75 X 17.6 INCHES

AUTHOR'S COLLECTION

The map by J. H. Young, from Mitchell's 1852 atlas, shows the boundaries of the recently completed Native American cessions of lands in Minnesota and indicates the vast stretch of land ready to be opened for settlement. A key of text reads: "Reference to the late Indian Treaties. The district bounded by a blue line, and lying on both sides of the Red River of the North, was acquired by the Chippewa Treaty of 1851. The district bounded by a red line, and lying between the Mississippi and the Sioux Rivers, was acquired by the Sioux Treaty of 1851."

Although no red and blue boundaries were hand colored on this particular copy of the map, it does show the boundaries of the Dakota (Sioux) land cession in Minnesota, which extends south out of hand-colored Minnesota into Iowa. Earlier maps indicated Native American ownership of land with words but not with actual boundaries. This map follows that convention in the unorganized counties west of the Sioux River, the western limit of the cession, and in the northern part of "Chippeway Country." The Sioux reserve extends along the Minnesota River from the northern edge of Big Stone Lake to just upstream from Prairie la Belleview. In addition, a New Sioux Agency is shown. The Dakota villages on the lower Minnesota River—frequently depicted on earlier maps as occupied by Black Dog, Shakopee, and Big Legs—as well as the missionary station at Traverse des Sioux, remain on the map though they were probably unoccupied by this time. The Indian-village name Kaposia appears on this map, but the name refers to the post office and incipient town that missionary Harriet Bishop founded on the site of Little Crow's former village. In addition to the label "Chippeway Country," the map shows an Indian agency for the Ojibwe (Chippewa) at Sand Lake. The villages at Fond du Lac and Red Lake are the

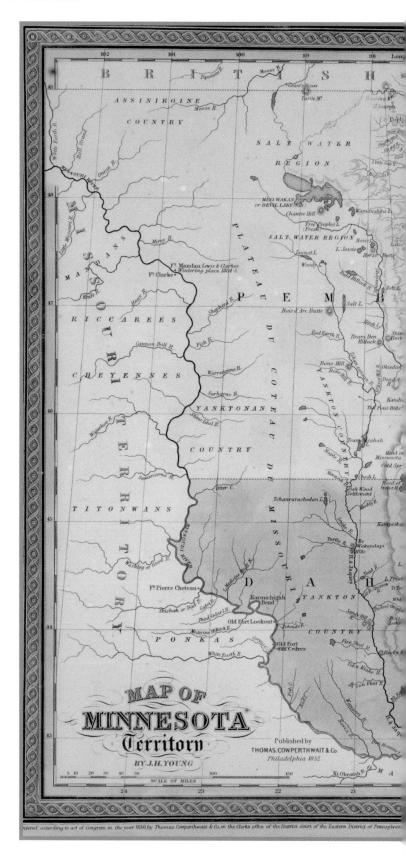

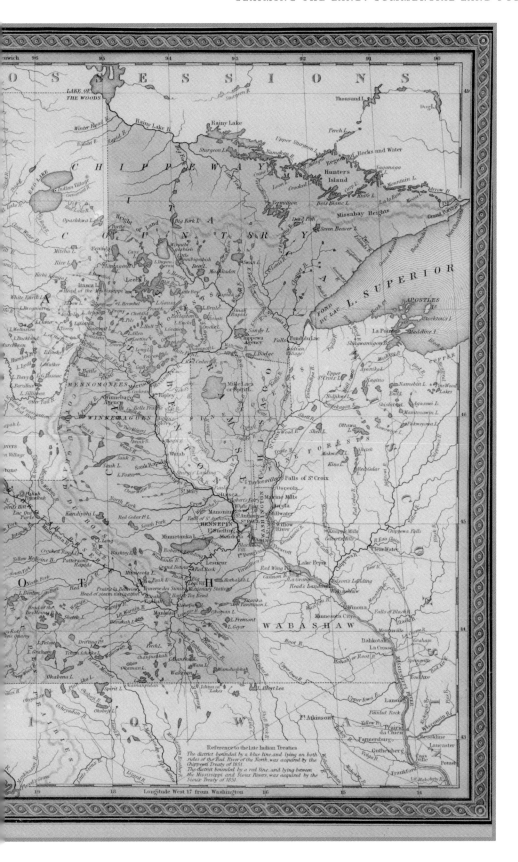

only two Ojibwe settlements that receive labels.

The map also indicates the lands of the Menomonies and Winnebagoes (Ho Chunk), with a reservation inside the northern portion of the Dakota cession. It sites the Winnebago Agency at Long Prairie. Under duress, the Winnebago had ceded their land in southern Wisconsin to the United States, after which a number of the tribe were removed to northern Iowa. In 1848, the government moved them again, to the Long Prairie site, and in 1855 they were settled on a reservation of eight townships in eastern Blue Earth and Steele (now Waseca) counties. After 1863, the government moved them out of Minnesota entirely.

This map shows the declining isolation of Fort Snelling and the growth of new Minnesota communities by its rendering of roads running along the Mississippi River and radiating out from the capital at St. Paul and the milling districts of the St. Croix. In addition, an improved road stretches northwestward from St. Paul, through St. Anthony and along the east bank of the Mississippi to a point above Fort Ripley. Here it joins the Red River Trail used by oxcarts going to and from Pembina.

The map's depiction of a fledgling transportation system and of land available for settlement suggests in geographical terms the imminent end to the frontier status of Minnesota Territory.

J. H. Young. *Map of Minnesota Territory,* no. 36, from *Mitchell's Universal Atlas* [?].

PHILADELPHIA: COWPERTHWAIT, DE SILVER AND BUTLER, 1854

14.5 X 17.5 INCHES

AUTHOR'S COLLECTION

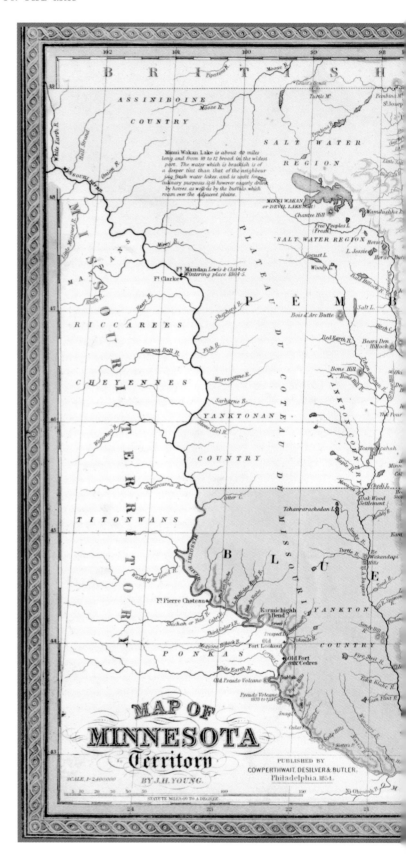

Like its immediate predecessor, Young's map from two years later suggests the quickly approaching end to Minnesota Territory's frontier status. The boundary of the Dakota cession of lands in Minnesota is colored in red, but the boundaries of the Ojibwe cession on the Red River are not yet colored. The map designer clearly wished to emphasize the expanse of land opened for settlement. Rather than the short legend from the previous map, the legend here reads:

Lands of the Dacota or Sioux Indians

By the treaties of Traverse des Sioux and Mendota, conducted in the year 1851, the Dakota or Sioux Indians ceded all their lands to the United States lying in Minnesota and Iowa between the Mississippi and Sioux rivers, and between Lat 42°37' N and Lat 4°10' N., extending N, and S. 338 miles and E. and W. for 100 to 250 miles. Area about 54,100 square miles or 35 million acres: comprising a region fully equal in extent to the States of Pennsylvania and New Jersey.

The Dakota Reserve, a tract of land on the headwaters of the Minnesota River 120 miles in length and 30 miles wide, has been appropriated by treaty for use of said Indians. The red line is the boundary of the Dacota cession.

On this 1854 map, the Dakota Reserve extends along the Minnesota River from northern Big Stone Lake to a place upstream from Prairie la Belleview, and the New Sioux Agency is shown. Now the Dakota villages along the lower Minnesota that were present on earlier maps (Black Dog, Shakopee, and Big Legs) and the Presbyterian Mission are absent. However, the Sisseton village on Lake Traverse, lying just inside the reservation's western border, remains in place. The placement of Ojibwe settlements and names have not been changed, and the Winnebago Agency at Long Prairie is named. The Menomonie Indians have no presence on this map.

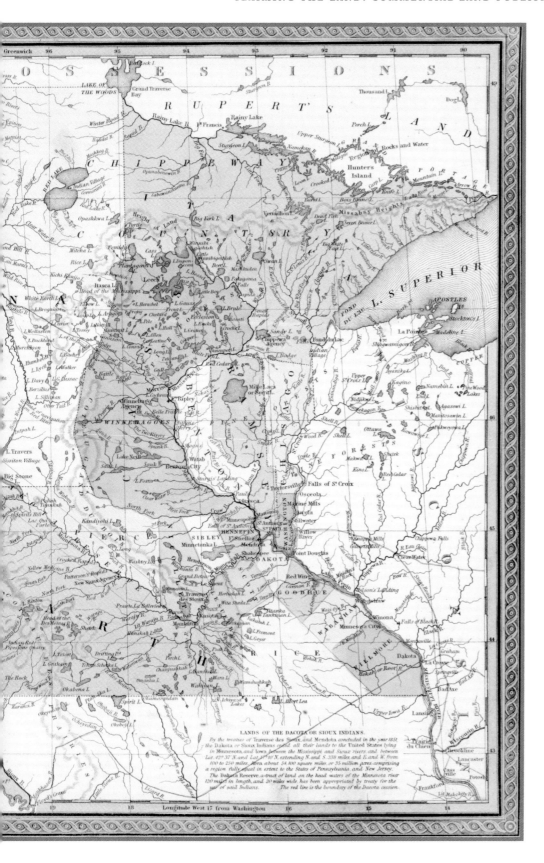

The pattern of roads that form the young territory's transportation system has not yet changed. The map indicates the head of navigation on the Red River of the north but does not mention the head of navigation on the Minnesota River. Port Charlotte and Grand Portage are also deleted from this edition of the map, but Fort Francis appears just across the Canadian Border on Rainy Lake.

Because of the territory's growing population, several new counties were created out of the original large, very sparsely settled "supercounties" — Wabashaw, Dakotah, Wahnata, Mahkahta, Pembina, Cass, and Itaska. On the north bank of the Minnesota River, Sibley, Nicollet, and Pierce counties replace the old Dakota and Wabasha counties. These counties are aligned with the river in a manner similar to the older French system of land holdings, wherein long, narrow lots afford more owners access to waterways. These counties would ultimately be realigned when the Public Land Survey had measured and formally recorded the property. Wabasha assumed something of its present size and shape, having lost its western territory to the new supercounty of Blue Earth and the other Mississippi River counties of Fillmore, Goodhue, and Dakota.

The river system and major topographic features on this map are little changed from earlier editions of the Cowperthwait maps. This text fills a blank area on the map in the area around Minni Wakan Lake, or Devil's Lake, which was little known: "Minni Wakan Lake is about 40 miles long and 10 to 12 broad in the widest part. The water which is brackish is of a deeper tint than that of the neighboring fresh water lakes and is unfit for culinary purposes, it is however eagerly drunk by horses, as well as the buffalo which roam over the adjacent plains."

These observations probably were added to make the distant plains more real to the map viewer, as well as to fill a geographically unknown area on the map.

Minnesota, no. 49,
from Colton's Atlas of the World.

NEW YORK: J. H. COLTON, 1855

14.5 X 17.5 INCHES

AUTHOR'S COLLECTION

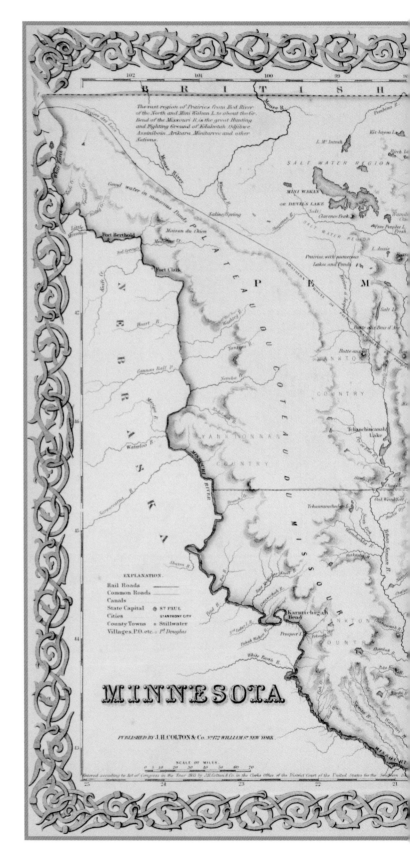

This 1855 map provides us with what is essentially a real estate developers' view of Minnesota Territory and its moneymaking potential. Minimizing the importance and presence of Native Americans, it removes many non-English place-names and focuses on the new transportation network and the area already being surveyed by the U.S. government.

Like earlier maps of the territory, this map reflects rapid changes in the area and bolsters Minnesota's case for statehood. The number of counties has doubled from that on Cowperthwait's map of a year earlier because the very active session of 1855 of the Territorial Legislature had authorized a dozen new counties. On later maps, many of the new counties would continue to change size, shape, and even location.

The counties of Goodhue, Wabasha, and Fillmore cease to be a strip focused on the Mississippi River and are transformed into the "squared-up" counties of Goodhue, Wabasha, Winona, Houston, Fillmore, and Olmsted. The 1854 dimensions of Rice County have been reduced, and Steele, Dodge, Freeborn, and Mower counties emerge. On both sides of the Minnesota River, the counties are also aligned with the new grid: Dakota, Scott, Le Sueur, new Blue Earth, Nicollet, Sibley, and McLeod. Now, several unorganized counties on the west-

ern side of the survey are visible. The old Blue Earth County's western portion is shown as an unnamed, unorganized area occupying the southwest sector of the territory. Pine County took a big bite from old Ramsey County, creating a gerrymander that so confused the map colorist that the county is shown in three different colors: yellow, green, and light pink. Thus began the shrinking of Ramsey County that culminated with it becoming the state's smallest county in area.

One of the puzzles of the map is the location of New Ulm and Brown County. Authorized in February 1855, the county was not organized until the legislative session of 1856. Apparently at that time it moved north from its position on this map to the south side of the Minnesota River. Its county seat, New Ulm, is shown on the Watonwan River rather than at its actual site on the banks of the Minnesota River several miles to the north. In similar fashion, the post office of High Forest shown in Mower County was called Pleasant Valley in 1855–56. The name had changed to High Forest in 1856, and in 1858 it was transferred to Goodhue County. Somehow the cartographers knew that the little post office of High Forest had changed its name in 1856, but they missed the fact that Brown County had been moved.

The territory's railroad system is represented very differently on this Colton map than on the Cow-

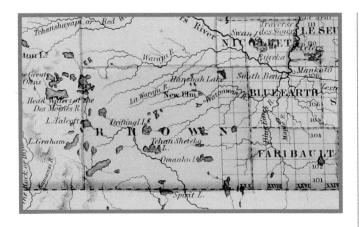

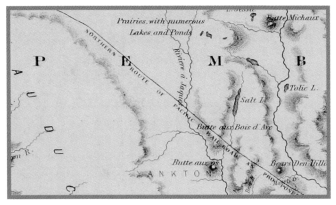

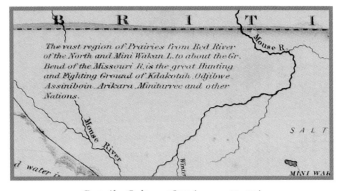

Details, Colton, 1855 (pages 52–53)

perthwait maps, again to reinforce the image of rapid development. Proposed railroads connect St. Paul to the east, crossing the river below Stillwater and bypassing both Clear Water (Eau Claire) and Chippewa Falls. Another railroad extends north from La Crosse along the east bank of the Mississippi. A third railroad runs from La Crosse through Winona and Rochester and on to Mankato. A bold line stretches across Pembina County, indicating the northern route of the Pacific Railway as proposed by Governor Isaac Stevens.

New wagon roads connect Rochester to Iowa. The road to the northeast reaches the St. Louis River. The territory's road network was developing with the crosscutting route north of Mille Lacs that connected Fond du Lac and the Mississippi River. The Red River Trail to Pembina and the wagon road north from Fort Ripley to Crow Wing have been removed, while the road to Long Prairie is indicated (although the Indian agency is gone). For some reason, Port Charlotte and Grand Portage appear on the map, but Fort William, a larger settlement, does not. Nonetheless, the message of the map is clear: this territory is being tightly bound to the United States, especially to St. Louis to the south and Chicago to the east.

The line of the Dakota land cession does not appear on this map nor do various names indicating ownership of land by Native Americans east of the James River. Several villages appear in the unorganized areas. The Red Lake Village and the Chippewa Agency at Crow Wing are named, but the Fond du Lac Village and the agency at Sandy Lake are not. The map notes the Sioux Agency on the Minnesota, but it neglects the territory reserved for the Dakota after 1851. A Presbyterian mission and trading post are indicated at Lac Qui Parle, and Sisseton and Yankton villages are shown on opposite sides of Lake Traverse. In contrast to the Cowperthwait maps, this Colton map does not show the southwestern pipestone quarry revered by Indians, but it adds the words "Great Oasis" for the headwaters of the Des Moines River.

Inexplicably, Colton replaces Cowperthwait's positive text box in the northwest corner of the territory west of Devil's Lake with this troubling message: "The vast region of prairies from Red River of the North and Mini Wakan Lake to about the great bend in the Missouri R. is the great Hunting and Fighting ground of Kdakotah, Odjibwe, Assiniboin, Arikara, Minitarree and other Nations."

The rendition of the physical landscape on this map is basically similar to that on earlier maps by Mitchell and Cowperthwait, but with some interesting additions. Hachuring shows the height of land between the Red Lakes and the drainage into the Rainy Lake system. This convention suggests a range of mountains in the midst of the big bog regions, when in fact the entire region, which is part of the bed of Glacial Lake Agassiz, is very nearly flat. Few place-names for physical features appear on this map, probably a good thing.

Minnesota, from *The World in Miniature, America.*

NEW ORLEANS: MORSE & GASTON AND A. B. GRISWOLD, 1857

7.5 X 6.25 INCHES

AUTHOR'S COLLECTION

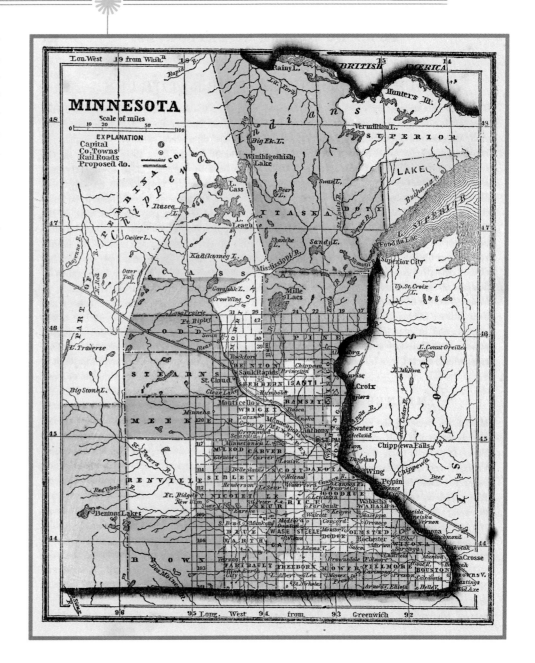

This time-stopping map shows counties in the northeastern Arrowhead Region that existed for only a matter of days. The map may be, in fact, the only one ever to delineate Doty and Superior counties in northeastern Minnesota. Published by partners in New Orleans that included descendants of Jedidiah Morse, the author of the first geography text about the United States, this small map portrays an attenuated version of Minnesota Territory with minimal detail.

Doty and Superior counties had the shortest lives of all the counties authorized within Minnesota. On February 20, 1855, the Territorial Legislature created Doty County in the area east and south of the St. Louis River; the rest of northeastern Minnesota became Superior County. Then, on March 3, 1855, for some unknown reasons, legislators changed Doty to Newton County and Superior to St. Louis County. Almost a year later, on March 1, 1856, St. Louis became Lake County and Newton became St. Louis County. The three counties shown on this map are a mysterious blend of the decisions of 1855 and 1856.

The impact of the Morse family publishing firm on geographic education in the United States is significant, but this map did not appear in any textbooks. It stands as the sole record of a long-ago event.

Colton's Minnesota and Dakota,
from *Colton's Atlas of the World.*

NEW YORK: J. H. COLTON, 1858[?]

12.25 X 15.25 INCHES

AUTHOR'S COLLECTION

This landmark map is the first to depict the new state of Minnesota, which was admitted into the Union as the thirty-second state on May 11, 1858. The controversial issue of whether Kansas would enter the Union as a free or slave state had delayed congressional passage of Minnesota's application for statehood for several months.

Colton and his contemporaries were easily able to produce the map, based on the older plate of the territory. They simply delineated, engraved, and colored the newly defined western border. A few other changes were made to the 1855 plate, but this map's primary purpose was to record for posterity the political decision associated with Minnesota taking its place in the Union, an event long predicted by the series of morphing territorial maps.

This map maintains the image of rapid settlement in the region and thus "justifies" Minnesota's admission into the Union as a free, nonslaveholding state. Figures from the 1857 census have not been compiled, but the 1860 census counted more than 170,000 non-Indians in the state, considerably more than the approximately 6,000 on the 1850 census. (The number of Indians in the early territorial era has been estimated at more than 30,000.) The number of counties has increased, and most have a proper county seat. Several counties in the western and northern parts of the state have yet to emerge, but the east central area has stabilized. The border with Canada is drawn along the forty-ninth parallel to the southwest corner of Lake of the Woods. The Northwest Angle is not yet on the map.

Although the copyright date of this map is given as 1855, it clearly dates from 1858 or 1859 because it shows the border between the state of Minnesota and Dakota Territory. It also has two Browne Counties. The remnant of the old Browne County would be organized as Watonwan in 1860.

The old supercounties are all but gone. The last of the southern supercounty southwest of the Minnesota River has been authorized as Browne County. The oddly shaped Toombs County and the large Pierce County were established in the west central part of the state, between the boundary of the Dakota cession of 1851 and the western boundary of the new state. In time, both of these counties disappear from the map. In 1857, a band of counties two-deep was organized along the Iowa border, from Faribault and Blue Earth west to Rock and Pipestone. Several counties were created out of Pine County, including Buchanan, which later would be absorbed by Carlton and Pine counties. Ramsey County lost territory to the new Anoka County, although further boundary adjustments would be made to the old northern section of Ramsey.

There are no changes in the representation of Native Americans in Minnesota, but statehood brought a very definite border between the Indian Countries in Dakota Territory and the more settled and organized lands in Minnesota. The route of the proposed railroad to the west remains, as does the description of the "hunting and fighting grounds" in the northwest.

Transportation and the Public Land Survey had expanded considerably. A railroad extends from the Twin Cities all the way to Sioux Falls; another line goes to the Canadian border. No distinction is made between existing and proposed roads, so once again we see evidence that the cartographer exaggerated development to promote settlement. New roads connect Rochester to the expanding agricultural settlements in the Iowa border counties.

Congress did not act to organize the westernmost land lopped off from Minnesota Territory until 1861. Thus, for three years it existed as unorganized federal land with no representation in Congress, despite persistent appeals from residents.

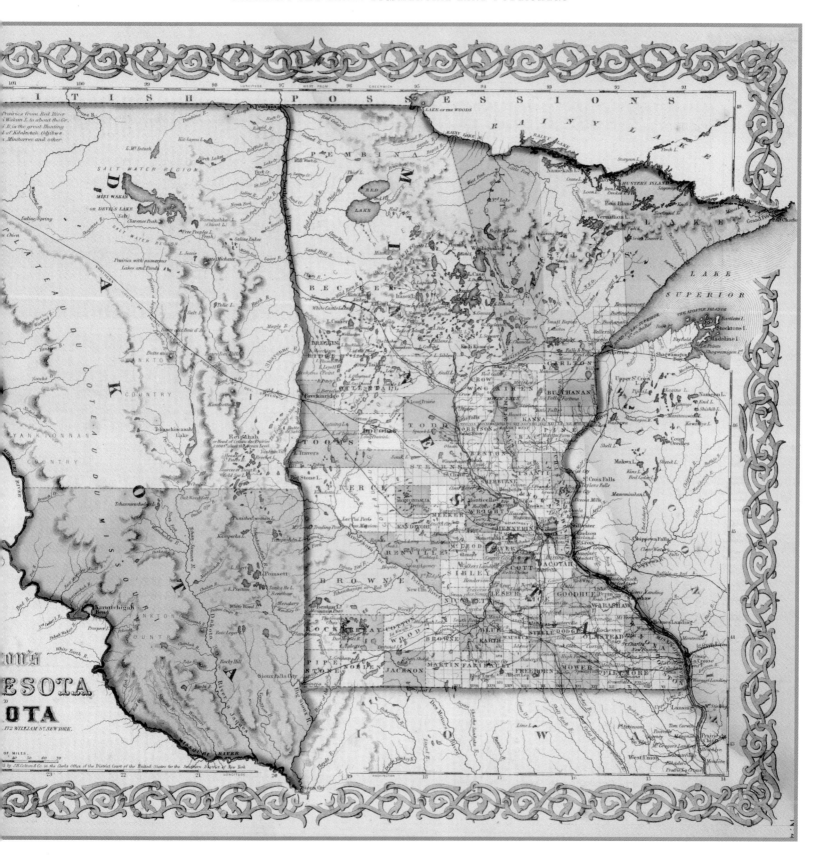

Johnson's Minnesota,
from *Johnson's New Illustrated (Steel Plate) Family Atlas.*

NEW YORK: A. L. JOHNSON, 1865

18 X 14 INCHES

AUTHOR'S COLLECTION

Six very large counties—Pembina, Polk, Cass, Itasca, St. Louis, and Lake—occupy the northern part of the state in this post–Civil War map showing the expansion of organized political units and settlements connected by roads and the beginnings of a railroad network. Many organized but unsurveyed counties remain in the southwest and in the Red River Valley, but the southeast quadrant is divided into townships and peppered with towns. The county of Andy Johnson is located on the western border, originally named in 1858 for Robert Toombs, a politician from Georgia. When Toombs became a leader of the Confederacy, county residents changed its name in 1863 to Andy Johnson County to honor Lincoln's vice president. Not pleased with the way that John-

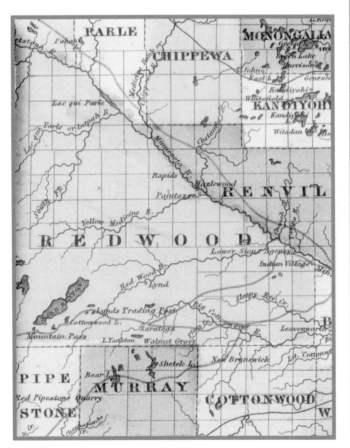

son governed, they changed the name again in 1868, this time to Wilkin County. Colonel Alexander Wilkin raised the first company of the First Minnesota Regiment in the Civil War. He died leading the Ninth Regiment in the Battle of Tupelo.

On the western edge of this map, Redwood County is in the process of being surveyed and consists of groups of unnamed and unorganized townships. A wagon road connecting the Lower Sioux Agency and village with Lynd's Trading Post and on to Sisseton lands in Dakota Territory winds through an unsurveyed region. This is not current information, however. Lynd lived among the Dakota with the intention of learning about them and writing a book, but he and most of his manuscript for the book perished in the attack on the Lower Sioux Agency in August 1862. The agency was abandoned after 1862, but the old map plates were not modified to record the passing of the Dakota.

The road system shown here is more accurate than in earlier, wildly optimistic maps of the territory, but the map still inaccurately reinforces the image of a rapidly developing state. Three proposed railroads are shown connecting St. Paul to Breckenridge to the west; one route connects the Twin Cities with Duluth via the central lakes region; and several roads run to the south and southwest. Because this map indicates both built and planned railroads, careful viewers could distinguish between fact and fancy. The lines that were eventually constructed connected St. Paul to St. Anthony and to Newport. Another independent road extends west from Winona toward Mankato but only reaches St. Charles. The great post–Civil War railroad building boom lay not too far ahead in the future.

Representation of Native American land ownership in the Johnson map differs considerably from that in earlier maps. Some villages and agencies are named, even though they no longer had occupants in 1867. The reservation boundaries are very indistinct, and the mapmaker used pictographic symbols and images of tipis to mark the villages, perhaps to suggest they could easily be removed.

To the east of the big bend in the Minnesota River, the agricultural lands have been surveyed and organized into townships. Labels for the burgeoning number of county seat towns compete with township names for the viewer's attention.

Place-names reveal a fascinating mixture of Native American, frontier, and new Americanized names. Chengwatana, an Ojibwe village at the place where the Snake River debouches from Cross Lake, for example, is clearly shown. This village was a long-established community that attracted fur traders during the fur trade era. It boasted the post office for Pine County until replaced by a practically adjacent town, Pine City, the English equivalent of Chengwatana.

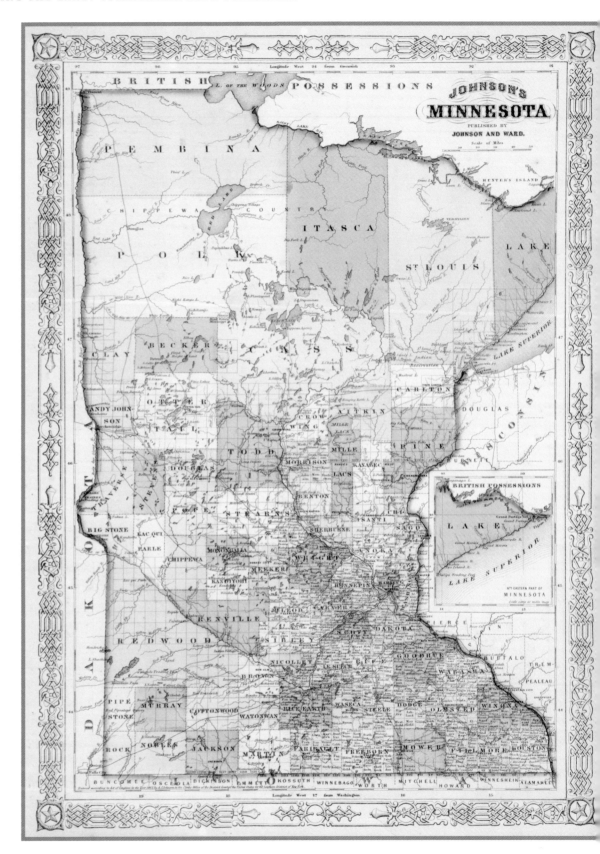

County Map of Minnesota, drawn and engraved by William H. Gamble, no. 81, from *Mitchell's New General Atlas.*

PHILADELPHIA: S. AUGUSTUS MITCHELL, 1877

15.25 X 12.25 INCHES

AUTHOR'S COLLECTION

This county map from 1877 continues the tradition of highlighting the state's expanding political organization, its growing web of towns connected by roads, and its developing railroad network. It shows a cartographic transition from maps designed to incorporate new places into the common body of geographic information to maps focused on the railroad network to meet the needs of immigrants, traveling salesmen, and freight shippers.

Based on an older map and still lacking the Northwest Angle, the 1877 map's key includes symbols for the state capi-

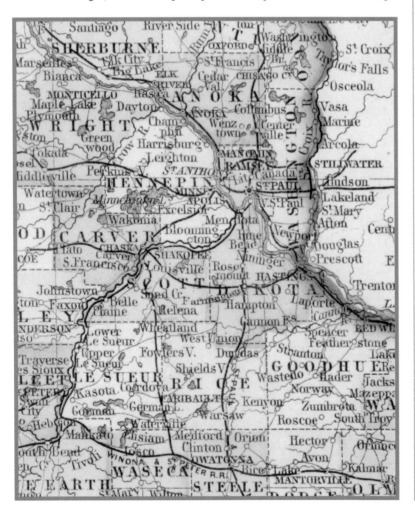

tal, county towns (seats), common towns, railroads, proposed railroads, and common roads. The map also marks the wagon road heading from Fort Ridgely to the Missouri River, the main military road to the west. The old supercounties are all but gone. Five very large counties—Pembina, Cass, Itasca, St. Louis, and Lake—make up the northern part of the state. No attempt is made to indicate the limits of the Public Land Survey, even though it was not completed in northern Minnesota. Redwood County is divided into Lincoln and Lyon counties. Lynd's Trading Post continues to be present, though the mapmaker has relocated the road that connected it to the east.

The road and rail networks are detailed, maintaining the image of a successfully developing state. Although the key has a symbol for proposed railroads, none are shown. The Mississippi River towns of Winona and La Crosse both have tracks connecting them to the network of towns and farms in southeastern Minnesota. However, the tracks are not completed through the Red River Valley to Canada. The road from Duluth up the North Shore ends at Saxton near the mouth of the Temperance River, so that the village at Grand Portage was accessible only by boat or primitive trail. For the first time, a mapmaker has indicated the extent of steamboat travel on the Mississippi, noting the "highest point reached by steamboat in 1858" on the Mississippi, close to modern Grand Rapids.

Mitchell shows no Native American reservations but includes a village and agency for the Dakota on the Minnesota River, even though that reserve no longer existed. The Ojibwe villages at Grand Portage, Gull Lake, and Fond du Lac have been removed. Leech Lake and Red Lake villages are shown, but the road connecting them has disappeared. Roads from the Chippewa Agency at Crow Wing north to the agency on Leech Lake are indicated.

The lakes and rivers are the same as shown on earlier maps by Mitchell and Cowperthwait. The watershed divide between the Gulf of Mexico and Hudson Bay is indicated, but the mapmaker does not indicate the triple divide near Hibbing and instead includes Lake Superior in the Gulf of Mexico's drainage system. He indicates the headwaters of the Mississippi and the north-flowing Red River of the North and gives an elevation of 1,608 feet for each. The pipestone quarry and the Coteau des Prairies have disappeared. This reflects the continuing de-emphasis of the physical environment and the increasing importance placed on the transportation system and the locations of growing cities.

Labels for the rapidly increasing number of county seats and other towns are peppered across the southern portion of the state, leaving no room for township names. A mixture of frontier and new Americanized names is still evident, but English names are the vast majority. Towns planned by speculators and railroad companies are strung along the tracks at short intervals. Their future seems comfortably assured.

4

OWNING THE LAND: COUNTY ATLASES

Title page, 1888 plat book

The mapping of Minnesota reached new heights after the Civil War with the private publishing of county atlases. These collections of local maps are a quintessential element of American cartography. They reflect the importance in the United States of local governments, property rights, the rule of contracts, and a thriving and vigorous real estate market.

Mediterranean and European concepts of land ownership and land taxation were founded on the ability to measure area and record rights to sections of the landscape. For many centuries, land surveyors had played a crucial role in the development of communities. To complement the official surveys of land parcels that were made and held in government vaults, detailed maps of land ownership and land boundaries, called cadastral maps, were also produced for use by real estate agents, tax assessors, and local government officials.

Midwestern publishers dominated the publication of county atlases beginning in the late 1860s. The several thousand county atlas or plat books published during that time vary tremendously in quality. Some are extraordinary works of cartographic and publishing craft with finely engraved plates, illustrations, and descriptive text.

Prosperous farm counties in the nineteenth-century Midwest formed a prime market area for the atlases, and the Midwest became one of the centers of this cartographic genre. Farmers were proud of their efforts to harness their land, grow crops, and raise animals. They built houses, purchased exciting farm machinery, and improved breeds of livestock. The existence of the Public Land Survey made mapping the Midwest relatively easy for publishers. Political boundaries conformed to the cadastral survey lines or to topographic features. Most farms were regularly shaped subdivisions of the townships based on the primary survey lines.

To make the maps, surveyors copied information about land holdings and ownership from official records in each county courthouse. With this base map in hand, they then drove the roads of the county, filling in prominent landscape or other features, including the locations of houses, commercial establishments, mines, forests, and anything else thought to be interesting. They frequently interviewed a few residents to make sure nothing of significance was missed. The science of cartography formed only a small part of the process; most of the time and costs involved in producing the maps involved canvassing for sales and the printing, binding, and delivery of the atlases.

Atlases were sold exclusively to "subscribers" within the county. Once company canvassers secured a few endorsements from prominent citizens, they would go door-to-door in the county selling a place in the future atlas to farmers, merchants, and others. Purchasers would sign a contract to buy the book. In addition to getting their own copy of the atlas showing the extent of their properties, they would also be listed in the back as subscribers. Once the sales contracts were in hand, another agent of the publishing company would contact the wealthy men in the county about having their personal biographies published in the atlas for the price of two and a half cents a word and a minimum of $10. For an additional fee, their portraits or images of their farms could be added.

Advances in printing technology made the publishing of county atlases quite profitable. A publisher could expect to receive about $18,000 to $25,000 in sales, half of which was profit. In order to achieve these returns, publishers focused on predominantly rural counties with populations over 10,000. Not everyone would order an atlas, but some would order multiple copies.

During the 1870s, Philadelphia became the leading publisher of county atlases, followed by Chicago and New York. Davenport, Iowa, St. Louis, Missouri, and San Francisco were secondary centers serving local markets. During the 1880s, Minneapolis joined the leaders of the industry. In the 1890s, Chicago replaced Philadelphia as the leader, and St. Paul–Minneapolis moved to third place. During the first decade of the twentieth century, St. Paul–Minneapolis displaced Philadelphia. In fact, in that decade, the Twin Cities produced twice the output of Philadelphia.

One of the most creative and motivated publishers of county atlases was Alfred Theodore Andreas. In 1867, an army comrade persuaded him to try his hand at canvassing door-to-door for county map publishers based in Geneva, Illinois. Working with old army friends, he started slowly, but in 1869 he came up with the brilliant idea of putting the county maps together into an atlas. Andreas soon dominated the business. He further expanded the compilation of maps into a county picture book, replete with illustrations of just about everything in the county, including houses, animals, and the properties of the middle class. He truly popularized the atlas. Along with popularity came prosperity. He bought out his partners and moved to Chicago. There he had access to specialized publishing firms located in the Lakeside Building, which housed all the component parts of atlas publishing under one roof.

During the three decades from 1890 to 1920, publishing county atlases was good business in Minnesota. Thirty-three different companies published 137 atlases of Minnesota during these years. Three publishers—Ogle & Co. of Chicago, Northwest Publishing Co. of Philadelphia, and Webb Publishing of St. Paul—published two-thirds of these atlases. Most of the atlases included illustrated advertisements, directories of residents, and lists of patrons.

After World War I, the atlas business changed. Modern utilitarian works produced by hundreds of small presses for local markets across the country have played an important role in the real estate business, providing cadastral information to the residents of counties and potential land purchasers. These atlases, however, often lack the craft and style of their predecessors.

No matter when or where an atlas was produced or what county was mapped, the purpose of the atlas was to promote local communities and businesses. Atlases praise the communities, local businesses, and the landscape, inviting the viewer to come, take up land, and join in the bright future promised in the maps.

Whitewater Township, from *Map of Winona County.*

CHICAGO: LYMAN G. BENNETT AND A. C. SMITH, 1867[?]

11.5 X 11 INCHES

MINNESOTA HISTORICAL SOCIETY

Atlases published by Lyman G. Bennett and A. C. Smith of Chicago were Minnesota's first atlases and among the earliest produced in the nation. They stand out because of the quality of the lithography and printing by Charles Shober & Co., proprietor of the Chicago Lithographing Company. These atlases have ornately lettered title pages and beautifully drawn maps showing both land-ownership patterns and elements of the physical landscape.

Bennett's Winona County atlas of 1867 is a beautiful work containing nineteen page-sized township maps and several engravings of major public buildings and sizeable homes. Its frontispiece is a foldout bird's-eye view of Winona from the Wisconsin side of the Mississippi River. The map of Whitewater Township illustrates the excellent cartography, lithography, and printing that characterize Bennett's atlases. The cartographer and engraver faced an interesting challenge in depicting the intensely stream-dissected coulee country of the "driftless," or unglaciated, area of southeastern Minnesota, with its steep bluff faces and flat-topped interfluves. But the team was certainly up to the task, and the map is an excellent example of hachuring.

The map also shows some of the problems associated with the Public Land Survey. Settlers were sometimes sold squares or rectangles of land that bore no relationship to the topography. This resulted in farms with two layers, valley bottoms and bluff tops, separated by a steep cliff. The 160-acre parcel owned by T. Kirk in section 29 and the parcel owned by D. McCarty in section 30 are excellent examples of the problem. The few roads built before the 1860s wound from the lower valleys to the uplands, where they merged into the system of roads that followed the township lines. Most of the houses were located near the roads. The farmers on the upland prairies secured supplies of fuel and building materials by purchasing woodlots such as the small parcels in section 20. For unknown reasons, real estate developers platted two towns—Beaver, with twenty blocks, in section 15 and Whitewater, with sixteen blocks, in section 27. These towns were quite close together, in fact, too close together to live up to the expectations of their founders. The population density was low, and the local market in the township was small. In addition, the towns were not easily reached from the productive agricultural areas on the upland.

Beaver suffered a particularly cruel fate. When settlers eventually stripped prairie sod from the upland and lumbered

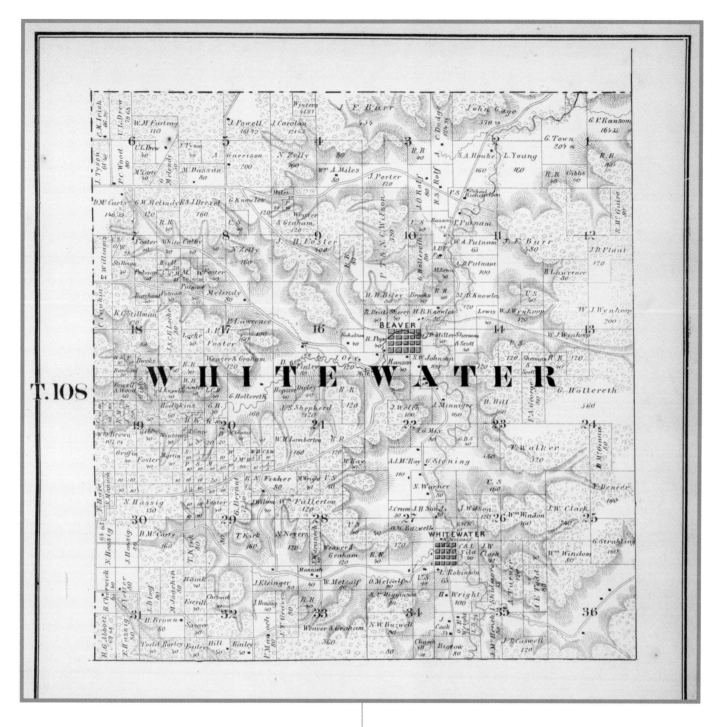

most of the forests from the hillside, rapid and spectacular erosion destroyed Beaver. The topsoil soon slipped away downstream and eventually buried the farms in the valley—and the village of Beaver as well. Over time, the erosion in the Whitewater Valley reached such a state that the area was removed from agriculture and converted into a state park.

Today, the river valley is managed for waterfowl and for many forms of outdoor recreation. The road that winds downstream from Whitewater through Beaver and leaves the township in section 1 is now State Highway 74, which is the last unpaved state highway in Minnesota. It is frequently closed due to flooding and snowfalls.

Shafer Township, from *Plat Book of Chisago County, Minnesota.*

MINNEAPOLIS: C. M. FOOTE AND CO., 1888

17.5 X 14.25 INCHES

MINNESOTA HISTORICAL SOCIETY

This early map of a portion of the St. Croix River Valley gives us a view of the river-oriented development of the region. Taylors Falls, clearly the dominant community in the township, was the port of entry for the largely Swedish immigrants that settled the Chisago lakes area in the western portion of the county.

The future of the St. Croix River and its valley was still undecided when this map was made. The rapids that formed the head of high water navigation are clearly depicted. A scheme to have the U.S. Corps of Engineers dig a shipping channel in the St. Croix and link the river to Lake Superior via a canal had been defeated, and a hydroelectric dam was built on the rapids. The dam interfered with the waterpower-based manufacturers downstream and of course ended the practice of floating logs to sawmills on the spring flood. The logging industry had passed its peak when this map was published, and we can see the farming communities emerging along the roads and railroads focused on Taylors Falls. The railroad actually comes to a dead end in the river town. The map indicates rural-based enterprises, such as the blacksmith's shop in the small hamlet of Shafer. Also visible are lots extending to the river from the bluff top. These would eventually become part of the state park established to protect the geologic features known as the Dalles of the St. Croix. The paving and upgrading of the highway made Interstate Park a favorite destination for tourists and day-trippers beginning in the 1920s, but in 1888, when this map was made, the township was a long way from urbanized areas. It was blessed with good soil and, as the map indicates, an abundance of flowing springs.

While the map is not ornate and the coloring is faint, it is an excellent example of the mass production techniques that characterized the early years of county atlas production. The map's story suggests the organization of the industry.

In 1872, George Warner, a well-established publisher of county maps in Iowa and Wisconsin, formed a partnership with Charles M. Foote, who apparently was new to mapmaking. Although they continued publishing county maps, they also began publishing bound atlases, beginning in 1877 with *The Atlas of Grant County, Wisconsin.* They located the headquarters of their firm in Minneapolis but had their lithography and other aspects of the production done in Philadelphia. Unlike many competitors, Warner and Foote did not publish profusely illustrated atlases. Their business partnership lasted until 1886, when Warner's name no longer appears. Thereafter, Foote ceased publishing county maps and focused entirely on atlases. He worked with Edwin C. Hood and John W. Henion, as well as independently, until 1899, when his son, Ernest B. Foote, took over. He published atlases until 1903. Through its quarter century of publishing, the firm maintained its focus on the Upper Midwest, publishing maps of Minnesota, Wisconsin, Iowa, and Michigan.

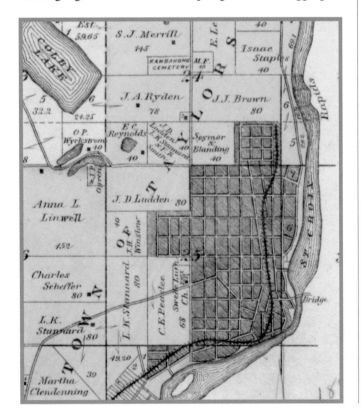

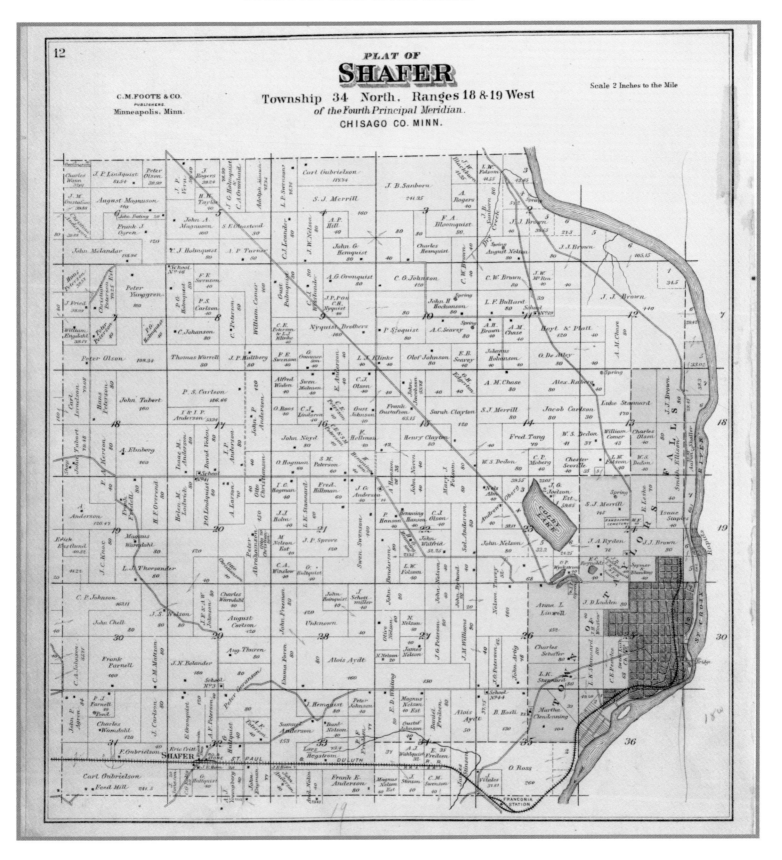

Swede Prairie Township, from *Farmers' Atlas & Directory of Yellow Medicine County, Minnesota.*

ST. PAUL: *FARMER MAGAZINE* (WEBB PUBLISHING), 1913

16 X 11.5 INCHES

MINNESOTA HISTORICAL SOCIETY

In 1882, Edward A. Webb, a newspaper reporter from the Fargo area, purchased a magazine and renamed it *The Farmer*. Written "strictly for and about farmers in Minnesota, North Dakota, and South Dakota," it was the first publication of its kind in the Midwest. This very successful business eventually became one of the largest printing and publishing organizations in the Upper Midwest. It served rural citizens living in townships such as this one in Minnesota's Yellow Medicine County.

A close look at Swede Prairie Township on the western edge of the state illustrates the tendency of European immigrants to congregate in homogeneous settlements, called "ethnic islands" or "communities of affinity." These clusters resulted from chain migrations, wherein immigrants were attracted to places where they already knew people from their home communities. Immigrants followed family members or heard about opportunities from letters sent back home. Immigrants preferred to settle among people who spoke their language, shared their culture, and were willing to help the newcomers.

In Swede Prairie, the names of the landowners and places are predominantly Scandinavian. Although it is impossible to distinguish Swedish from Norwegian names, it seems safe to assume that this is a Swedish community because it is centered around the Swedish Lutheran Church. A Norwegian Lutheran Church is located just 5 miles to the northeast, in adjacent Norman Township. The directory lists each household, giving the names of the head of household and spouse and their children. As a special feature, it indicates how long the heads of the households had lived in the state. A healthy handful of men named Ole lived in the township, but, strangely, not one Sven. Schools are scattered across the area. The schools were needed because, according to the directory, families were large, especially the family of Peter and Emma Peterson, who had lived for twenty-one years on 160 acres in section 24 with their thirteen children. (In contrast, J. A. and Belinda Olson farmed 160 acres in section 7 with their three children and had lived in the state for only one year.) Scandinavians brought their political economy with them, and in the northeast corner of the township is sited the Spring Creek Cooperative Creamery.

In 1913, farmers in this area were on the cusp of the prosperous few years that resulted from the demand for wheat and foodstuffs produced for World War I. The era of semi-subsistence farming was just about over. Life on the family farms of Swede Prairie in 1913 was filled with hard, labor-intensive work. In a generation or so, the increased mechanization of agriculture and rural electrification would greatly diminish the need for farm labor and begin a process of increasing farm size.

In addition to ethnic information, the Swede Prairie map provides some information about the land market. Two financial institutions, the Iowa Land Company and the Globe Land and Loan Company, owned parcels of land. We can only guess at the history of these parcels. Perhaps the ownership of these two parcels resulted from foreclosures—certainly not the last farm loan foreclosures in the region.

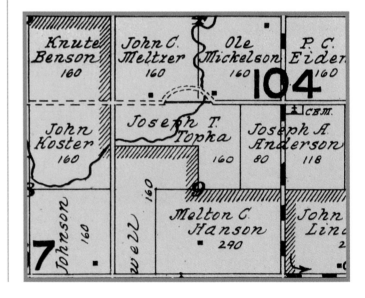

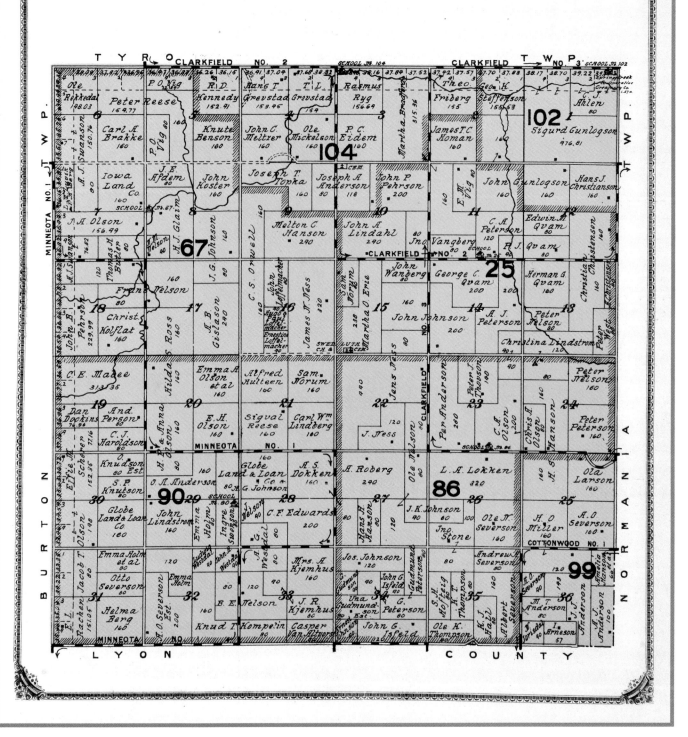

Cottage Grove Township, from *Plat Book of Washington County.*

MINNEAPOLIS: NORTHWEST PUBLISHING, 1901

18.5 X 15 INCHES

MINNESOTA HISTORICAL SOCIETY

Cottage Grove Township, southeast of St. Paul on the Mississippi River, was first settled in 1844. Sited beyond the exurban zone at the turn of the twentieth century, it was totally transformed into a suburban landscape in the 1950s and 1960s by the Panorama City addition in 1955, the Orrin Thompson Grove additions of 1957 through 1959, and later additions in the 1960s.

Visible in this 1901 plat map is a portion of the string of towns that were platted along the railroad tracks leading out from St. Paul through St. Paul Park to Summerfield and St. Paul Park Acre Lots toward the independent hamlets of Langdon and Cottage Grove. Early in the territorial era, farmers had settled Washington County. Unlike the central and western portions of the state, however, this area was not an ethnic enclave. The names of landholders have a British Isles tone, but it is not possible to know if the settlers came from states farther east or from overseas. Major roads, which predate the Public Land Survey and wend their way across the bluffs and the river, cut many farms into two pieces. Despite the presence of the larger stock farms, the farms are on average smaller than those in the prime agricultural areas farther south. These small farms were primarily dairy farms providing fresh milk for markets in the Twin Cities.

The map shows Cedarhurst in section 3, the summerhouse of St. Paul attorney Cordenio A. Severance and his wife, Mary A. Harriman Severance. Just beyond the urban fringe, this property began as a post–Civil War farm and provided an elite retreat for this St. Paul attorney's family and friends. The Severances built a twenty-six-room Neoclassical Revival–style mansion, thanks to the talents of Cass Gilbert, Minnesota's most famous architect of the period. Presidents Theodore Roosevelt and Warren Harding and other dignitaries visited the estate during the 1920s and 1930s. Severance and Frank B. Kellogg were partners in one of St. Paul's most prestigious law firms. Kellogg was elected to the U.S. Senate and later was appointed to a variety of positions, including that of secretary of state. The connection between the two men may have led to the local myth that the house played a role in the negotiations of the Kellogg-Briand Peace Pact of 1928 that earned Kellogg the Nobel Peace Prize of 1929.

During the Great Depression, this exurban landscape attracted a much more humble population: workers from the cities who engaged in a form of going back to the land. In those difficult times, people supplemented their income or mitigated their unemployment by moving in with farmer relatives. Others rented small farms and worked the land while maintaining their town jobs. Some commuted to St. Paul by rail from the Langdon station.

The Northwest Publishing Company that produced this map was based in Philadelphia from 1892 to 1899 and in Minneapolis from 1900 to 1910. The firm published 120 plat books. All included a county map, individual plats for the townships and towns, and maps of the United States and Minnesota. The maps seem to have been lithographically printed and hand colored. Like most other plat books, they contain no filler illustrations. The term "plat book" would gradually replace "county atlas" in the midwestern vernacular. Descendants of this book were increasingly simple and less expensive but maintained their focus on property holders.

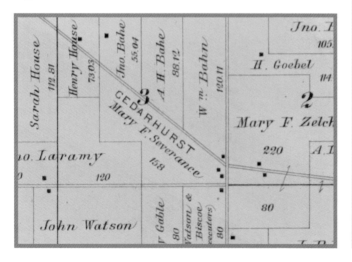

PLAT OF COTTAGE GROVE

Scale 2 Inches to the Mile.   Townships 26 & 27 North. Range 21 West.   of the 4th Principal Meridian.

5 | MAPPING THE STATE: THE ANDREAS ILLUSTRATED ATLAS

Detail, Andreas, 1874 (pages 80–81)

In 1874, map publisher Alfred Theodore Andreas issued the fruits of several years of organizationally challenging and innovative work. His *Illustrated Historical Atlas of Minnesota,* an extension and elaboration of the popular county atlas, reflected the rising interest in local geography and the flush of agricultural prosperity in the Midwest.

Andreas was a very successful map publisher, but his competition was strong, especially after the most inhabited and therefore most profitable counties were rapidly mapped. Andreas, a quintessential marketer, realized he needed a new product to keep his atlas publishing business profitable. County atlases were largely focused on farmers and depicted their land, but state atlases were intended to appeal to the entire population. More illustrations of businesses, government buildings, churches, and factories would be included in his unique product. Some additional maps of the United States based on the 1870 census would be added to increase its value to would-be land and business buyers.

Andreas selected the state of Minnesota because it had not yet attracted significant attention from other county atlas publishers. Mapping the state was a big job; distances were long and the transportation system was spotty. His production team, based in St. Paul, eventually included 108 people, who took the project from beginning to end in two years. In addition to door-to-door salesmen, Andreas began to use the mass media to advertise the forthcoming atlas, and he paid newspapers to publish stories he wrote about it. Andreas seems to have completely understood the need to combine massive advertising and public relations campaigns with a mass-produced product to successfully manufacture and sell his new atlas.

Because the Minnesota atlas would be a collection of maps of counties and towns, he could not publish maps at a scale large enough to show the names of all the property holders. He solved this problem by printing only the names of subscribers who prepurchased the atlas. His canvassers found a ready market and were able to sign up 12,000 subscribers at a minimum fee of $15. Because the northern part of the state was lightly populated and difficult to canvass, the entire area is relegated to one double-page spread in the atlas. A handful of subscribers are shown around Detroit Lakes, but Duluth has only one. However, one enterprising salesman managed to get H. H. Helm, a resident of Superior, Wisconsin, to subscribe, and thus a piece of Wisconsin is included on the map. It is estimated that one out of every seven Minnesota households subscribed to the atlas, but the rate was surely higher in the prosperous agricultural counties of the south central part of the state. The atlas contains 350 picture views, 340 portraits, and 219 biographies. The venture was expected to gross $250,000, and Andreas expected to make a profit of $50,000. Unfortunately, the Panic of 1873 intervened, and Andreas could not collect from all of his subscribers.

Andreas wrote in his preface: "We cannot but congratulate ourselves on the sense of relief we feel from the responsibilities which have weighed up on us during the preparation of a work so unique and voluminous. Few persons without actual experience can comprehend the details of such a work, its cost and the care and pains necessary to bring it to completion." He went on to write a little piece titled "What It Takes to Make a State Atlas":

The lithographic engraving and lithographic printing were done by the Chicago Lithographing Company, Chas Shober & Co. . . . To give an idea of the magnitude of the task, we will say that if the labor had been confined to one man and one press, it would have taken the man thirteen years and the press two years. . . . If one man had set the type for this work and arranged the forms and done the electrotyping, it would have taken him two years and a half and one press about six months to print it. The coloring is the work of Warner & Beers and will speak for itself as to the careful manner in which they have accomplished their work. If one person had done all the coloring, it would have taken him forty-five years.

The printing of Andreas's atlas of Minnesota required 70 tons of paper, and the case binding required 17 tons of cardboard. A ton each of leather, cloth, and other materials were also used. The 10,000 atlases were bound at the rate of 400 per day. There is little reason to doubt Andreas's assertion that "there is not another city in the Union where a work of this magnitude could have been gotten out in any reasonable length of time."

While the cartography of the maps in the Andreas atlas is less than brilliant, the atlas is a remarkable document of the cultural landscape, made at a time when large sections of the state were not yet developed. The lithography is certainly equal to the standards of the time. Illustrations exhibit the usual features of the county atlas. They contain all sorts of people dressed up on their way to someplace, but few people are working. Women push baby carriages in the road or drive fancy buggies. Men are shown in a variety of ways, some driving new machinery. Frequently pastures extend up to houses so that more animals can be shown. On occasion, livestock fill pastures with large, ungated openings that make the livestock easier to see. Not only do we find farmers showing off their best animals and crops, but the proprietor of a pump and pipe store can be seen on the sidewalk in front of his shop holding a pump for all to admire. The illustrations range in size from 2-inch squares up to the full-page view of the 1,800-acre Lake Elysian Stock Farm of C. A. De Graf, reported to be one of the many described as the "largest and finest farms in the state." The *Atlas* also includes three gate-folded, full-color, bird's-eye views of three river cities, St. Paul, Minneapolis, and Winona.

Andreas's *Illustrated Historical Atlas of Minnesota* represents a milestone in both the development of American cartography and the image of Minnesota. For the first time, the grandeur of the entire state and the details of locales were portrayed in one volume. This book was not just about where farmers owned land; it was about the productivity of agriculture, the prosperity of commerce, and the advancement of the professions. Homes ranging from the grandest mansion to the unadorned farmhouse are proudly displayed, each making a contribution to the development of the state. Andreas also published an abridged history of the state and counties in the atlas.

Unfortunately, publication of new state atlases involved high overhead and sizeable capital. Andreas rolled the profits from the *Illustrated Historical Atlas of Minnesota* into a much-grander *Atlas of Iowa,* and before that was finished he began an Indiana atlas. Because the sales of the atlases were so closely tied to the prosperity of farmers, the vagaries of weather and agricultural markets played havoc with the subscriptions and collections for the atlases. The problems of obtaining capital and the uncertain financial conditions of the time forced him to declare bankruptcy in December 1876. Andreas paid his creditors sixty cents on the dollar and closed his atlas publishing company. His career in map and atlas publishing had blazed like a meteor and then extinguished.

All told, Andreas published twenty-three lavish and popular county atlases and three state atlases of unprecedented size and quality. When economic conditions changed, the era of atlas publishing ended, never to be revived. After the end of his cartographic ventures, he started a new company that successfully published local histories, and later he worked with the panorama shows popular in Chicago at that time. He died at the age of sixty-one, leaving his wife to survive on a Civil War widow's pension.

The Minnesota atlas provides us with a unique view of the evolving state. Although not all the residents and landowners are shown, it holds a wealth of detail. Using the county maps with topographic maps, we can generate a view of the Minnesota agricultural frontier and understand the pride that the people took in their accomplishment. Although the territory of the Dakota was greatly diminished by the 1870s, the Ojibwe are still clearly a visible presence in the northern sections of the state, with their reservations, villages, and agencies all marked. The sample of maps that follows reflects some of the rich diversity of information found in the Andreas atlas.

Anoka County,
from *Illustrated Historical Atlas of Minnesota.*

CHICAGO: A. T. ANDREAS, 1874

17.5 X 14 INCHES

MINNESOTA HISTORICAL SOCIETY

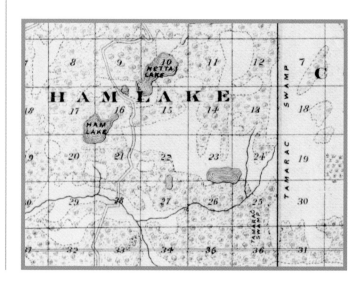

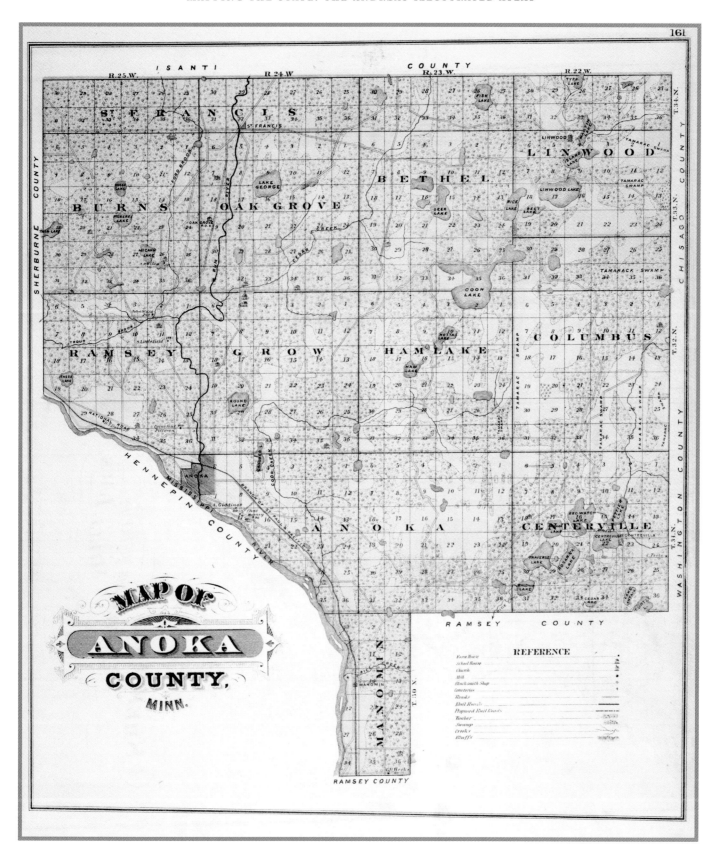

This map of Anoka County, northwest of St. Paul, was probably intended to be the first map in the state atlas, because its unusual shape made it possible to include a map legend on the printing plate. No other maps have a key of this kind. When the atlas was finally printed, however, the Anoka map was placed well back in the atlas, so that its key is not particularly helpful. Because the county was sparsely populated in the 1870s, it apparently had only a half-dozen or so subscribers who signed up to purchase the expensive atlas. Most of the eastern section of Ham Lake and all of Columbia and Lincoln townships are shown covered with tamarack swamps. A segment of the old government road between Fort Ripley and Fort Snelling appears, as do rails heading westward toward Puget Sound. Today the county is home to expanding Twin Cities suburbs, including Blaine, Ramsey, and Spring Lake Park.

Counties of Pine, Kanabec, Isanti, and Chisago, from Illustrated Historical Atlas of Minnesota.

CHICAGO: A. T. ANDREAS, 1874

17.5 X 14 INCHES

MINNESOTA HISTORICAL SOCIETY

This fascinating multicounty map shows the pattern of forests as they change with latitude in these east central Minnesota counties. In the south are clusters of oak trees growing in the grasslands near North Branch and Chisago. Farther north and west is a zone of red oaks. Maple Ridge and Rock Creek townships are the southern limit of deciduous pines. In the north are the Snake River pineries.

In the southern part of the map, a network of roads, towns, and farms extends through the mixed forest. Although a railroad runs through the counties on its way to Duluth, most of the towns are not on the rail line. Instead, networks of roads focus on Cambridge, Isanti, Sunrise, and other communities, because railroad branch lines have not yet been built. The only indication of the ethnicity of the inhabitants is Swede Mill near Stanchfield.

The most valuable land depicted on this map is the area north of Pine City. Although the farmers were not eager to pay to have their properties included, the lumbermen were. As a result, seventeen lumber camps are identified, spreading across Pine County. These settlements did not last long, however, for in a few years the forests shown here became stump lands, as the trees were cut for lumber to build farms and cities across the Upper Midwest and beyond.

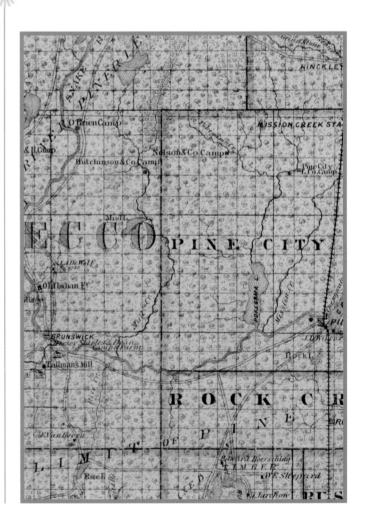

Dakota County,
from *Illustrated Historical Atlas of Minnesota.*

CHICAGO: A. T. ANDREAS, 1874

17.5 X 14 INCHES

MINNESOTA HISTORICAL SOCIETY

This atlas map provides an interesting view of the free-running Mississippi River along the northern boundary of Dakota County. Visible are numerous large islands that occupied the river before it was impounded behind a series of dams. Even though the map does not show local relief or contours, topography can be inferred from the pattern of woods and lakes found in the northern section of the county. This is an area of low moraines caused by glacier deposits. To the south are the flat and fertile plains that attracted many farmers who could afford to subscribe to the atlas and thus be recorded for posterity.

The county's road system focuses on the river ports of Hastings, West St. Paul, and Mendota. Roads clearly follow the path of least resistance as they wind through the landscape and down the ravines into the valley of the Mississippi. The limited number of rail lines focus either on Hastings or Mendota and intersect in Rosemount. Several towns are in the process of developing to serve neighboring farmers. A cheese factory on the Cannon River near present-day Randolph indicates the mixed farming practiced in the area. In years to come, this area would become the primary source of milk for the Twin Cities. Cheese making, which required less efficient connections to its markets, would become more important in towns farther away and less well connected to the Twin Cities.

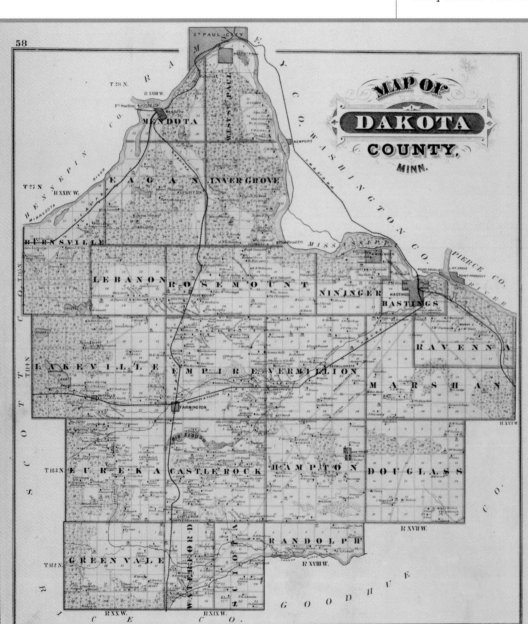

Northern Minnesota, from *Illustrated Historical Atlas of Minnesota.*

CHICAGO: A. T. ANDREAS, 1874

17.5 X 38 INCHES

MINNESOTA HISTORICAL SOCIETY

The northernmost part of the state contained very few people willing to purchase the atlas for $15, thereby assuring that their property would be on the maps, and Andreas gave this large area just two sheets in the atlas. Much of the northwestern part of the state was in fact imperfectly known to the mapmakers. This map is the only one in the atlas with a large blank area. In other areas lacking specific details, mapmakers filled the spaces with generalized forest patterns such as those used in the northeastern section. Of course, the landscape of northwestern Minnesota was not actually unknown, only unmapped.

Andreas's cartographers have captured the advancing Public Land Survey, expanding railroad trackage, and the beginnings of non-Indian settlement in the region. The surveyors were directed to work in places that had potential economic value. Therefore, the entire North Shore was surveyed before the rest of the northeast, and all of the Red River Valley was surveyed before the rest of the northwest. The land along the railroad already produced timber and was expected to become agricultural. Thus, cartographers measured and described it with an eye to future sales. Logs were also harvested along the North Shore, and the mouths of the small rivers were expected to become ports in the near future. The bog region around Red Lake and the northern forests were left by surveyors until last, because they were unlikely to be developed.

The Northern Pacific tracks extend westward from Duluth to the Red River Valley. Approximately seventeen towns had been platted by this time along the tracks between Duluth and Moorhead. The tracks to Winnipeg were not yet finished, and only Crookston had been platted in the northern portion of the Red River Valley. Because the historic Red River Trail followed bluffs on the west side of the river, no oxcart or wagon roads are shown in the Minnesota portion of the valley.

There was, however, significant real estate speculation in the valley and timberlands. The map shows a J. B. Bottineau owning over 9,000 acres in the valley north of Crookston and 1,440 acres of timberland between Aitkin and Leech Lake. He may have been the lawyer from Minneapolis known as John B. Bottineau who was involved in the litigation of Indian treaties in Minnesota and other states. In addition to the towns along the railroad, a smattering of lumber camps along the wagon roads built into the forests north of the railroad connect the Indian agencies at the reservations (with the exception of Nett Lake) to the railroad. The major Ojibwe reservations are shown: White Earth, Red Lake, Leech Lake/Cass Lake, Nett Lake, and Fond du Lac. Nett Lake is shown as a reservation, but it was not created by treaty, and at the time this map was produced and printed, members of the band had not agreed to settlement terms.

The map also shows a very interesting "Indian Trail" going southwest by west from the western shore of Leech Lake along the Crow Wing chain of lakes toward modern Park Rapids. Today's State Highway 34 follows this same route. This is the only Indian trail shown on any of the Andreas maps, and we do not know why, of all the thousands of trails in the state, this dead-end trail was the one included. Other roads include the route up the North Shore from Duluth to Beaver Bay, where a small community engaged in lumbering and later fishing.

The only settlement north of the rail corridor indicated on the map is a town called Vermilion, located approximately on the site of the modern town of Tower. This settlement and the road leading to it resulted from a small and unsuccessful gold rush to northern Minnesota in 1865 and 1866. The settlement shown is most likely a representation of a gold miners' camp built by members of the Mutual Protection Gold Miners Company (the Mutuals) in the late winter and spring of 1865. This miners' camp was never incorporated but was named Winston City, after one of the leaders of the company.

The company of miners carved the 80 miles of road connecting the mining area with Duluth with great difficulty between late December and early March. They followed an older trail that needed widening, clearing, and bridging. The forest was thick, the bogs deep, the hills steep, the snow three feet deep, and the temperature cold. Only the lure of gold could persuade men to undertake such an enterprise.

The road did its job, and by April Winston City boasted two stores, a group of bars, fourteen houses, a city hall, and headquarters for the Mutual Company. About 300 miners were active in the area, with more on the way. When the road thawed and the bogs reappeared, the miners experienced shortages of food and had to spend time fishing and hunting. The miners were undeterred, however, and soon about 500 men were engaged in prospecting and crushing the very hard rock. A mill and a crusher were brought in and set up across the lake from Winston City. The summer was one of hard work and very little gold. With the onset of winter, most of the miners left. A few returned the following summer, but the rush was over and a real town never materialized. The area had plenty of wealth in the form of iron ore, but its development required industrial-scale mining and extensive financing. The road cut by the gold miners was improved and made passable year-round in 1869 by the iron mining interests. This route is called the Vermilion Trail today and is on the alignment of modern County Highway 4.

George Shultz, a government surveyor from Duluth, recognized iron ore in local rock and surveyed a town site in 1882. It was incorporated in 1883 to serve the locations where the Minnesota Iron Mining Company built workers' housing close to the mines. It was named in honor of the Philadelphia financier Charlemagne Tower, who was a major shareholder in the mining company and the Northern Pacific Railroad. Nothing is known about why the Andreas cartographers labeled Winston City "Vermilion," but Tower is credited with being the oldest city in the Arrowhead Region and on the Iron Range.

The map shows the northern frontier of European settlement and the contact zone of whites and Native Americans. It also shows what the railroad promoters and timber men envisioned as an American Main Street linking the two coasts in a great transcontinental transportation system. Competition from other lines and cities, financial setbacks, and limitations of the environment would prevent this stretch of rails from providing the key link in the system. However, a decade after the map was published, the beginnings of the iron ore industry would create an entirely different cultural and economic landscape in northern Minnesota.

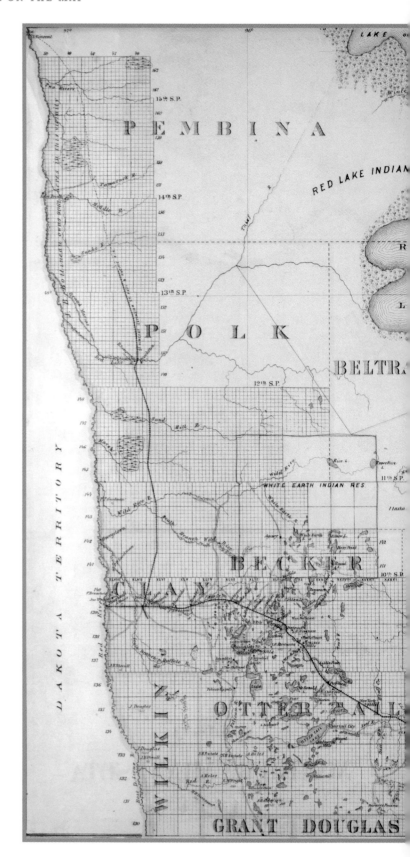

6 CITY PLATS AND MAPS

Detail, Duluth Street Railway Co, 1917 (page 103)

Establishing a network of urban settlements to provide services to farm families was crucial to successful settlement of the American Midwest. In most of the world, rural landscapes consist of farmers settled in clusters or hamlets and surrounded by their fields. In the United States, however, where land was more available and cheaper, farmers lived in a more dispersed pattern, in keeping with the Jeffersonian ideal. Because hamlets did not exist that could supply the goods and services farmers required, new towns were needed all across the agricultural frontier. Thus, the American landscape became clearly divided between urban and rural space.

In a symbiotic relationship, farmers needed towns to process, market, and ship their products and to supply the goods and services that could not be produced on the farm. Merchants and service providers in the towns, on the other hand, needed the farmers' business to provide an economic base. In addition, farmers needed a governmental system to uphold their claims of land ownership. Thus, although many popular images of the frontier focus on the far-ranging mountain man and fur trader, the frontier of agriculture was also an urban frontier. Cities were not literally in the wilderness, but the wilderness was frequently visible from their outskirts.

Towns in Minnesota developed in two phases. Some towns began prior to the Civil War, when transportation was dominated by river travel and only a few trails and roads penetrated the land between the rivers. Others developed along with the railroad network, when roads and trails were still primitive.

Most of the towns founded in the first era were the work of real estate speculators operating from eastern cities. They typically organized into small syndicates of investors and expected to establish thriving river ports on the Mississippi, St. Croix, and Minnesota rivers. Short wagon roads radiated out from these towns as farmers began to till the land. Small crossroads settlements and trading posts were created during this phase, but they were few and struggled for commercial life. The most successful of these early towns were those anchored by mills set up to saw the logs harvested in the virgin forests of the Mississippi River's watershed.

Minnesota did not attract large, organized groups of settlers who bought lands together and organized early towns that reflected their special philosophies or religions. New Ulm, founded by German Turners, may be the sole example.

After the Civil War, railroad companies, either directly or through their own special town site development corporations, platted towns on the land they owned along their tracks. These companies were thinking in monopolistic terms and wanted to have a set of towns that would complement one another and minimize competition.

The mining towns of the Iron Range were founded during this same railroad era. However, the urbanization of the range varied significantly from the pattern in the southern part of the state, because it included both incorporated and unincorporated communities, called "locations," set up for workers on land owned by the mining companies. Range communities were connected by railroads to the great ore docks in the ports of Duluth, Superior, and Two Harbors. Only a few other timber, mining, and manufacturing companies built and maintained towns for their employees, all in the northern section of the state.

Almost every settled township in Minnesota boasted a town where two roads crossed. Bigger communities contained a post office, an ethnic church, a general store, and perhaps a mill. By 1870, there were about 1,000 small towns in the state, all aspiring to greatness.

The U.S. legal system contains almost no regulations governing the platting of towns. Motivated individuals could plat a town site at any time and on any place they legally owned. No standards specified sizes or designs, although the plat needed legal property descriptions, streets, and blocks of lots. Accordingly, in this laissez-faire system, risk was borne by land speculators, who surveyed the site, drew up a plat, and registered the plat with the county registrar of deeds, and by the sometimes-gullible people who purchased the lots.

People had some expectations of what a town should be and how it should appear, but they tolerated a wide range of designs and sizes. Under this set of rules, however, the likelihood of failure or underperformance was high. Developers did everything in their power to ensure success, but their efforts were limited by the town's location and design. Issues of lot size and orientation, width of roads, efficiency of intersections, block size, provision of public land, and access to rivers, railroads, and roads could determine the appeal and success of a plat. To minimize the uncertain impact of town design on sales, developers frequently used designs from successful towns elsewhere.

Land speculators in Minnesota created towns for the purpose of stimulating commerce and developing real estate. They sited these new towns on either a river or a railroad and planned for a commercialized Main Street, leaving notions about a central village green and a courthouse square behind in New England and Pennsylvania. In only a few cases were public spaces considered central to Minnesota town design.

While developers wrote little about the thinking that lay behind their specific town plans, it seems likely that the county surveyors who approved the plans for registration were very influential in the process. The plats provided a spatial framework that set up a legal property holding system that promoted transportation, residential harmony, and commerce. These functions could occur in the context of a wide range of geometries, but Minnesota towns tended to fall into three patterns: they were grids that matched the lines of the Public Land Survey, they were oriented to railroad tracks, or they ran along the banks of rivers.

As the state's agricultural population waxed and then waned over the course of a century and a half, so did the fortunes of the farm trade centers, regardless of their design. Some towns grew large while others stagnated and declined. The rule of thumb that railroad towns were separated by the distance a horse-drawn wagon could travel in a day was abandoned after the arrival of cars and trucks. The development of rural free mail delivery made it possible for customers to order directly from catalogs and bypass the nearby town's merchants. These factors, plus the attraction of life in the big city, produced a great rural-to-urban migration that dramatically reduced the need for urban functions in small towns. Places with special features were able to grow, but most towns did not expand beyond their original plats.

Red Wing, Goodhue County,
from *Illustrated Historical Atlas of Minnesota.*

CHICAGO: A. T. ANDREAS, 1874

14 X 17.5 INCHES

MINNESOTA HISTORICAL SOCIETY

Red Wing is probably the oldest continuously inhabited place in Minnesota. Evidence of human habitation dates from 500 to 1,000 years ago, when people of the Mississippian culture lived in the area. Sited on a wide fertile plain that is the bed of a former channel of the Mississippi River, Red Wing was called Remnicha (wood, hills, and water) by the Dakota who lived there.

Following the 1851 Treaties of Traverse des Sioux and Mendota, which ceded to the government large portions of the Dakota lands in Minnesota, Red Wing was surveyed and platted in 1853. (The Dakota went to reservations or dispersed, although some groups later returned to nearby Prairie Island.) Settlement of Red Wing by whites burgeoned during the spring of 1854. Within a few years, the village was transformed into a busy river port and trading center that met the needs of the surrounding agricultural villages.

Red Wing's location at the edge of southern Minnesota's wheat region, along with access to more distant markets via the Mississippi, destined the city for early success. In 1874, some 2.5 million bushels of wheat were shipped from Red Wing, which made it the largest primary wheat market in the world at the time. The early prosperity of the city enabled businessmen involved in the wheat industry to diversify and support a group of important industries, in particular the manufacturing of shoes and pottery.

The initial plat mapping of Red Wing aligned streets and lots along the riverfront, but prosperity soon demanded a larger town. When W. P. Campbell mapped the city in 1874, it was already sprawling along the valley and up the cliffs. The curve in the river enabled town designers to be creative, and they used the opportunity to develop Central Park, between Fourth and Fifth streets and East and West avenues. The southern end of present-day Central Park was the original location of Hamline University, which opened to students in 1856 but relocated to St. Paul in 1869. This central community space attracted monumental churches and eventually a grand

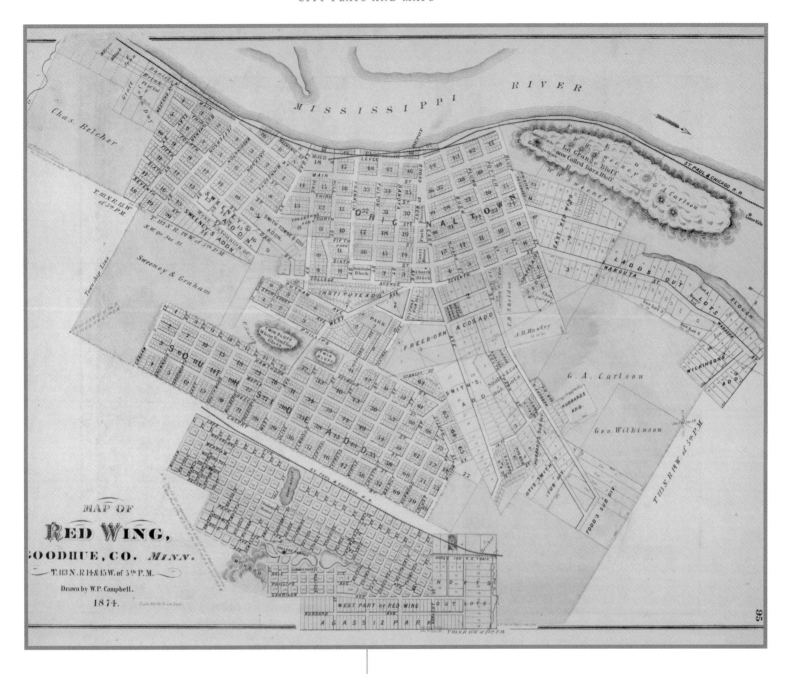

courthouse and provided a setting for popular community festivals.

During the 1870s, the town's riverfront was taken over by railroad tracks and grain milling industries, and the commercial core of the city developed parallel to it. The twists in the original plan required many revisions. Subsequently, highway and bridge building connected Red Wing to the newer transportation systems.

This 1874 map clearly indicates Barn Bluff, the city's tall rock guardian. A landmark for river travelers, with its massive size and height, 334 feet above the river, it never fails to impress viewers. Early explorers named it Mt. La Grange, because they thought it resembled a large barn, thus resulting in what is probably the least grandiose and romantic name given any landmark in the state. If it was the site of the grave of Dakota leaders, as the map indicates, they have a spectacular resting place.

Preston, Fillmore County,
from *Illustrated Historical Atlas of Minnesota.*

CHICAGO: A. T. ANDREAS, 1874

7 X 14 INCHES

MINNESOTA HISTORICAL SOCIETY

Becoming a county seat was the surest way for a new town to succeed. County governments provided jobs and required printing offices for official documents. The county's favored printer could also publish a newspaper and run advertisements for local merchants. Citizens coming to town on legal business also shopped and sought entertainments.

Political tensions ran high when decisions were being made on the location of county seats because there was little to differentiate one tiny town from the next when counties were first organized. In 1860, a dispute over the best location for Freeborn County's seat developed between Albert Lea and Itasca. When the question was put to a referendum, both towns bought votes. Trusting their future to a horse purchased surreptitiously in Iowa, the backers of Itasca challenged the champion racer of Albert Lea, a horse named Old Tom. The race was organized, with heavy wagering on both sides, and the winners intended to use the proceeds to ensure that their town won the county seat. What the Itasca backers did not know was that individuals from Albert Lea held a clandestine heat between the two horses and learned that Old Tom was faster. As a result, they were happy to match all the pre-race money Itasca could raise. When Old Tom won the race, the backers of Albert Lea used the money effectively to win the county seat referendum by a margin of nearly four to one.

While being a county seat was clearly important to a town's future development, the men platting towns in Minnesota did not choose to follow the central courthouse square plan that was seen in nearby states such as Iowa and in eastern states. That traditional plan calls for an entire block in the center of the community to be set aside for the courthouse, edged on all sides with the main commercial streets. Minnesota, in fact, has only one such organized county seat, Fillmore County's Preston.

John Kaercher, a miller, founded the village in 1855 to take advantage of the water-power site on the south branch of the Root River. It is impossible to know what he was thinking about the future of the village, but we can assume that he had great expectations. He not only platted the town in this unique manner, ignoring the town commercial core at his mill site, but was able to have the township, which was organized after the village was platted, take the same name.

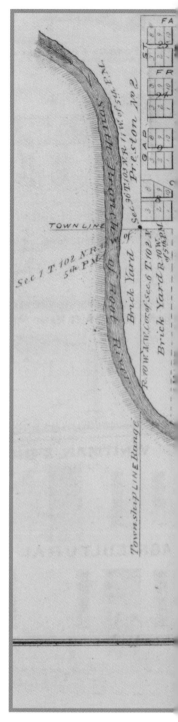

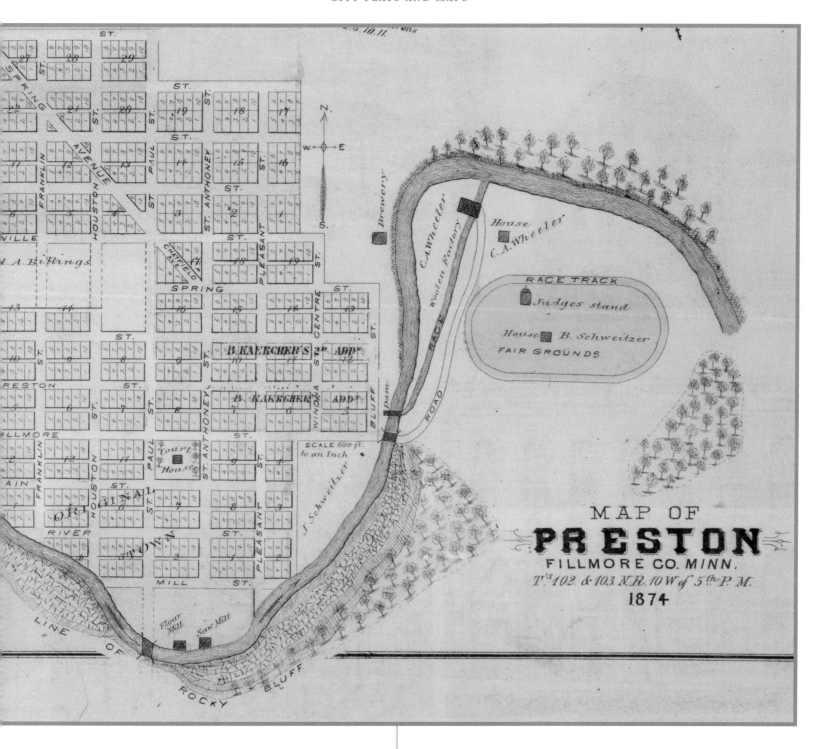

Unfortunately for Preston, when the Minnesota Southern Railway made its way up the Root River into the productive agricultural regions of the state, it bypassed Preston. In 1874, when this map was published, several nearby towns were larger. The local topography constricts the site, resulting in a plan that is not a perfect example of the central courthouse plan, but it is the closest approximation of a town planned for politics that exists in Minnesota.

Rochester, Olmsted County, from Illustrated Historical Atlas of Minnesota.

CHICAGO: A. T. ANDREAS, 1874

17.5 X 14 INCHES

MINNESOTA HISTORICAL SOCIETY

In 1854, the Minnesota legislature authorized two roads to Dubuque, Iowa. The main one entered the state via southeastern Fillmore County and then proceeded northward through Rochester and on to St. Paul. By the time this route was officially recognized, it was already an immigrant highway. In the spring of 1854, a caravan of 30 wagons and 150 cattle owned by Norwegians was reported on the road. At one time during that summer, approximately 200 wagons were on the road. That same year a village was staked out at the site of Rochester. The rude wagon road was improved for stagecoach service by men working for the company of Martin O. Walker, which operated stage lines in several midwestern states. Once the road had been somewhat improved, a coach drawn by four horse teams, which were changed every 15 miles or so, could make the trip from Dubuque to St. Paul in four days. In 1855, a major east-west stagecoach road going from Winona to St. Peter crossed the Walker road at Rochester, thereby making the village the most accessible interior village in southern Minnesota. The wagon roads and the availability of waterpower spurred the growth of the fortunate city, and by 1858 the settlement had been successfully incorporated.

The wagon road era of settlement in Minnesota was relatively short. By 1868, the Winona and St. Peter Railroad reached Rochester, and the economy boomed, with the export of huge amounts of wheat and flour from the very productive hinterland. By 1870, Rochester was the fourth-largest city in the state. In the next decade, the Chicago Great Western built a north-south railroad route through the city.

Rochester's long-term growth began with the arrival of the Mayo family in 1863, whose pioneering medical practices made the town a destination for medical treatment and diagnosis. Two Mayo sons attended medical school and began a general practice. Following a devastating tornado in 1883, Mother Alfred, from Rochester's convent, proposed the building of St. Mary's Hospital. In the age before antiseptics were in use, hospitals were not seen as desirable services to have in a town, but the Mayo brothers helped lead the profession toward sterile practices. When St. Mary's Hospital opened its operating room in 1889, the Mayos began pioneering new surgical procedures. They employed other doctors, to create a skilled medical group that could do research, diagnose illnesses, and provide treatment in one location. Mayo Clinic grew steadily into a world-renowned medical center.

Rochester in 1867

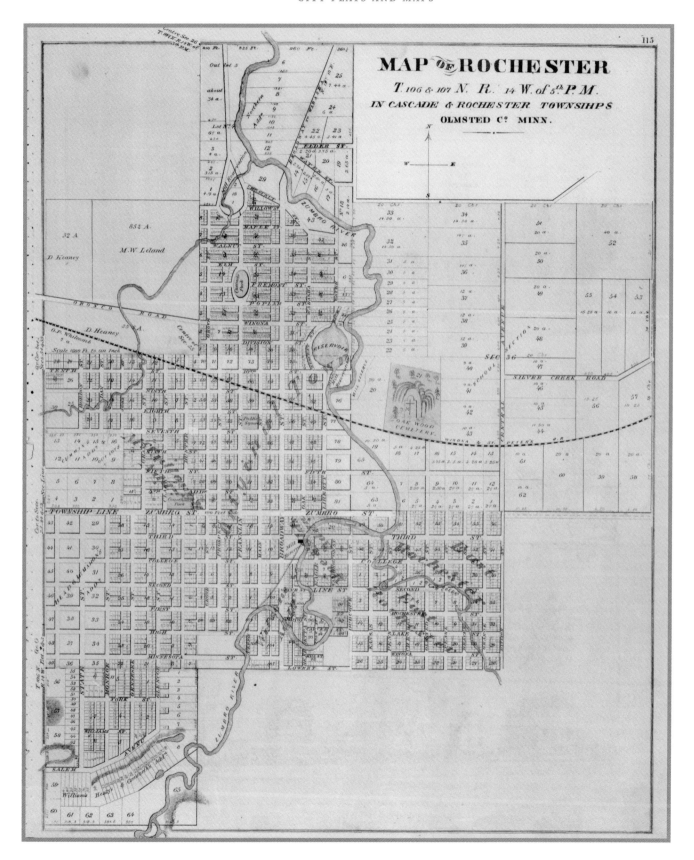

Heinze Bros. *Map of City of Brainerd, Crow Wing County, State of Minnesota.*

MINNEAPOLIS: JOHNSON, SMITH & HARRISON, 1883

42 X 30 INCHES

MINNESOTA HISTORICAL SOCIETY

The original plan for the city of Brainerd demonstrates the basic elements of all rail-oriented settlement plans. Because Brainerd was a major command-and-control point for the railroad system, the plan has additional features that illustrate typical industrial landscapes, as well as the transportation landscapes of railroading.

In typical railroad towns, land along the tracks was held by the company. The company then leased the land to commercial operations that needed direct access to a railroad siding, including grain elevators, lumberyards, coal yards, and livestock chutes. Railroad companies were perfectly willing to lease land to any and all of the grain dealers who operated in Minnesota, companies that were competing with each other. Therefore, in highly productive agriculture regions, towns frequently boasted several companies' grain elevators next to the depot and freight warehouse.

Because towns served to link large urban markets in the East and local farmers in the Midwest, and because towns were usually evenly spaced along the tracks, it was expected that each town would have the same set of services lined up on Main Street. Banks provided for financial needs, and hardware and implement dealers provided the wherewithal to make the rapidly mechanizing farms work. Groceries, dry goods stores, and drugstores provided goods for the homes. Some towns had a newspaper and print shop. In places with larger rural populations, there were competing enterprises supplying these services. In dairying areas, a creamery, frequently operated by a farmers' cooperative, could be found at the edge of the commercial district. The railroad hired land agents for each town, and these men competed with each other to entice the right mix of commercial ventures to establish themselves in their towns.

Plans for railroad towns were variations of two basic ideas. In the early railroad town plan, railroad tracks divided the town into halves, with business streets on both sides of the tracks. It soon became apparent, however, that trains steaming through the business district were disruptive and dangerous. Therefore, the more popular plan had the main street perpendicular to, or occasionally at another angle to, the tracks. Because railroad tracks were not laid according to the rectangular survey, these towns stand out within the grid of farm fields and township roads. Railroad town plans had public spaces for schools, parks, and cemeteries at the edges of the original plat, which were on land that the railroad company could not sell quickly. Growing towns were expanded by additions of various sizes. Some of these broke with the initial plat and followed the rectangular survey.

Builders of the Northern Pacific Railroad decided that the route would cross the Mississippi River a short distance upstream from the older settlement of Crow Wing on its way from Duluth to Tacoma, Washington. Once the crossing point was determined, a settlement sprang up to house workers leveling the grade and cutting wood for the ties. By 1871, the first construction trains had arrived in Brainerd, named after Ann Eliza Brainerd Smith, wife of the president of the Northern Pacific, John Gregory Smith. She may not have felt honored at first, because the early town was home base for 8,000 lumberjacks, who supported thirty-six saloons and auxiliary establishments. The hastily planned town site was quickly developed with simple log and frame buildings. During the initial boom years, tents and other temporary structures joined them. Most buildings clustered around Front Street, parallel to and south of the tracks. After construction of the Northern Pacific tracks to Brainerd was complete and crews moved westward, the railroad set up its headquarters and offices in town, as well as its main rail yard. These new economic undertakings drew many supporting industries, such as railroad tie making, and new residents poured in.

Brainerd's city plan was more elaborate than most because the city was designed to house the railroad headquarters. The plan is of the early variety; that is, the tracks split the town, resulting in two parallel commercial streets. In addition, a Main Street running north and south bisected the commercial and transportation area. To the north, a huge area encompassing four complete blocks was set aside for Gregory Park. The plan sets aside several blocks for churches, the courthouse, and schools. To the east of town, a huge shop and marshaling yard would be built to manufacture and repair rolling stock and manufacture ties, but this area was not part of the plat.

Brainerd also became the headquarters and railhead for the Brainerd Northern International Railroad, which ran to International Falls. This railroad, developed in the 1890s, was intended to serve the growing timber and paper industries developing at the end of the century. As in many other successful towns, Brainerd residents and investors created new economic activities, and today this area is one of the state's premier recreation centers. Because tourists and resort visitors come by car or plane, the old railroad corridor has many vacancies and relict land uses.

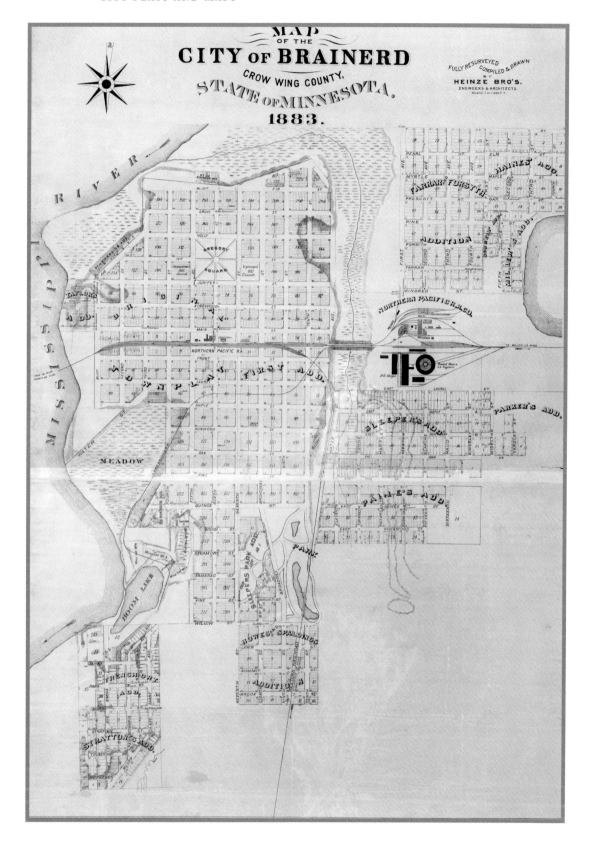

Litchfield, Meeker County,
from *Illustrated Historical Atlas of Minnesota.*

CHICAGO: A. T. ANDREAS, 1874

17.5 X 8 INCHES

MINNESOTA HISTORICAL SOCIETY

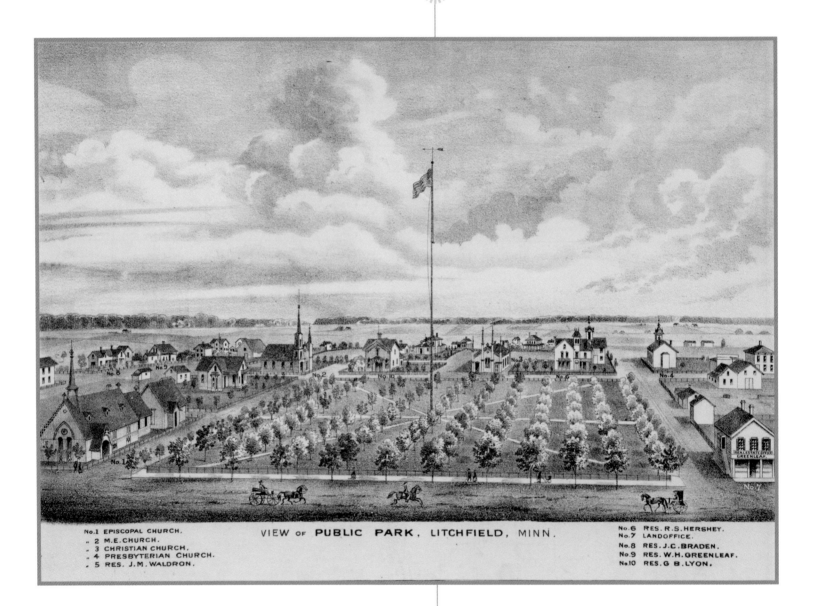

The arrival of railroads strongly influenced how Minnesota's cities grew and prospered. The process is demonstrated by the rise-and-fall relationship between two west central Minnesota villages. Meeker County's Forest City was platted in 1856 as a waterpower site on the North Fork of the Crow River. This hamlet boasted a sawmill, two flour mills, the first post office in Meeker County, and the government land office; it served as the county seat.

After the Civil War, in 1869, the St. Paul and Pacific Railroad platted the village of Litchfield, four miles away in prime

agricultural land, just a few miles beyond the edge of the Big Woods. Named after a family of financiers and railroad executives, Litchfield took over the status of county seat when Forest City was completely bypassed by the railroad. The village of Forest City failed to thrive and was eventually unincorporated.

Meanwhile, Litchfield prospered, and five years after its original plat had been filed, there were several suburban additions. This 1874 plat map showing the town layout reveals Minnesota's common perpendicular Main Street design. In this case, Sibley Street crosses the tracks next to the depot. As in most county seats in towns developed by railroads, the courthouse is outside the town center, and the public squares are dispersed to serve different sections of the town.

Hibbing Quadrangle, Minnesota.

7.5-MINUTE SERIES TOPOGRAPHIC, U.S. DEPARTMENT OF INTERIOR GEOLOGICAL SURVEY, 1957

27 X 22 INCHES

AUTHOR'S COLLECTION

This Geological Survey map illustrates two unique types of urban settlements that are typical of Minnesota's Iron Range: mining towns and mining "locations," or company towns. Prominent are the mining town of Hibbing and six nearby locations: Leetonia, Mahoning, Pool, Mitchell, Redore, and Scranton. Most Iron Range urban places began as mining locations, where mining companies built housing adjacent to their mines to accommodate workers. Typically a mining company held title to the land and either built small houses for renting to the miners or leased lots at modest costs on which miners might build houses. Miners and their families were responsible for maintaining the houses and grounds. Locations all provided the same basic function, but their forms varied extensively.

Hibbing, laid out by German miner Frank Hibbing in 1892, is a perfect example of another type of settlement, the mining-frontier boomtown. In 1895, Hibbing had almost five men for

every woman in town, and the population grew at a very rapid rate, from 2,000 people in 1910 to 15,000 inhabitants in 1920.

Mining companies were able to sell their land for residential and commercial development without passing along the mineral rights to that land, and if the company discovered ore underneath towns, it could force the settlements to move. At first miners and their families did not greatly protest, but when the Oliver Mining Company wanted to get at the rich ore under Hibbing, a tense struggle between the company and the city erupted. Finally, in 1918, the company successfully forced the town to move one mile south of the original site. In response, Hibbing's citizen leaders taxed the mines as much as they could and thus financed the construction of Hibbing's $3 million public high school. The great amount of space given the high school and community college in the new town plan indicates the value that the people of Hibbing placed on education.

By the time this map was made in the 1950s, demands for armaments and matériel for the Allied forces in two world wars had exhausted the reserves of high-grade iron ore, and Hibbing's population had started to decline. The map dramatically depicts Hibbing's perilous site amid a huge open pit mine, tailing ponds, railroad tracks, and piles of unusable overburden piled in man-made hills labeled mine dumps.

In 1964, the passage of an amendment to the Minnesota State Constitution encouraged the development of taconite mining and processing, which dominates the landscape today. As new ore bodies were developed, moving days returned to the range. Several of the locations that appear on this map, including Mahoning, Pool, Mitchell, and Redore, have disappeared in the expansion of the Hull-Rust Mine, the world's largest open-pit facility. Modern mining operations require a smaller labor force than in the past, and improvements in roads and automobiles make it unnecessary for workers to live adjacent to the mines and taconite plants.

It is hard to overemphasize the power of iron mining in the culture and economy of this northeastern urban frontier. In addition to the cities, the great holes in the surface, and the man-made hills, even the land survey was impacted. Highly magnetic ore deflected the compasses of the surveyors, causing odd angles to appear in the property lines and road alignments on the map. ❈

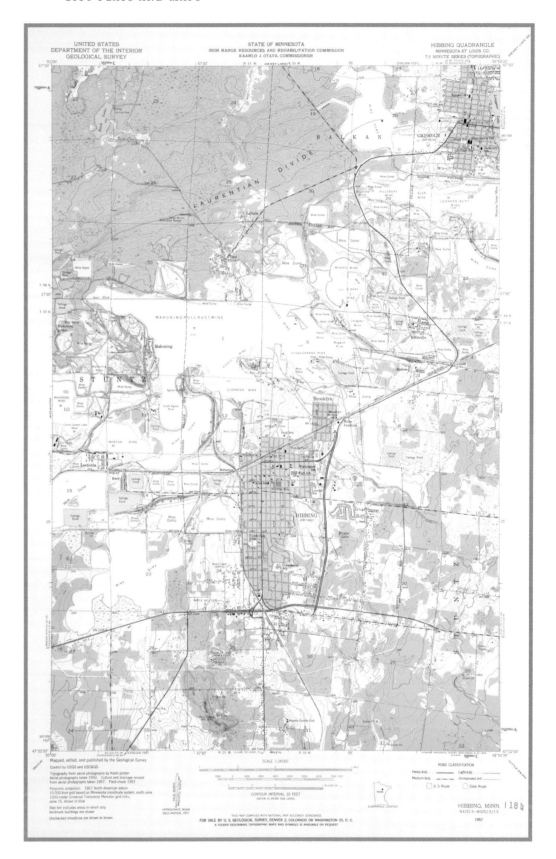

New Ulm, Brown County, Minnesota,
manuscript copy of Christian Prignitz's 1858 plat map, certified accurate by Fred Pfaender, son of Wilhelm Pfaender and Brown County register of deeds in 1896.

27 X 44 INCHES

BROWN COUNTY RECORDER'S OFFICE

Minnesota has only one community, New Ulm, that was founded upon utopian community ideals, and its story is unique. In the 1850s Ferdinand Beinhorn arrived in the United States with the intention of establishing a German colony. He traveled to Chicago, founded the Chicago Land Verein (Land Society), and, by advertising in the German-language papers, quickly attracted 800 members ready to move west. In 1855, the society claimed enough Minnesota land for a town and nearby agricultural areas, and within a year New Ulm, named for the city of Ulm in Germany from which many original settlers emigrated, quickly became a viable town and the Brown County seat.

In the same year of 1855, Wilhelm Pfaender, a leader of the Turner movement in Cincinnati, arrived in St. Paul to establish a German American colony. He called on his fellow Turners and other Germans "to unite for the establishment of a settlement, which aside from the material welfare, would also offer the advantage that the insane, degrading, mortifying attempts of our Anglo-American taskmasters to restrict us could not operate, that, in a word we have the opportunities to enjoy unstintedly the rights guaranteed to us by the Constitution of the United States, and become happy and blessed after our own fashion" (Tyler, 27).

In 1857, Pfaender's followers joined forces with the Chicago Land Society to create the German Land Association of Minnesota. It hired Christian Prignitz to create a plat for a greatly enlarged New Ulm settlement that would be able to accommodate the new migrants from Cincinnati, but when the Panic of 1857 severely strained the company's financial resources, it disbanded. In 1862, the organization donated its remaining land and cash to the New Ulm school district with the proviso that the Bible or other religious books not be part of the public school curriculum. Other association land was reserved for the fire station and hospital. Pfaender had successfully created a community for free thinkers and liberals.

The most dramatic feature of the Pfaender town plan is the presence of a garden belt around the city lots. Members of the German Land Association were to receive both city lots and large garden lots of four acres. This mixture of town and country is reminiscent of older agricultural hamlets in the German lands of northern Europe, where farmers lived in clusters surrounded by fields. A four-acre garden plot would not make a family self-sufficient but it would provide a quantity of fresh produce in the summer, potatoes and other root crops for the winter, and a supply of cabbage for each family's own recipe for sauerkraut. The town's developers, well aware of the high prices for food in the nineteenth century, wished to offer residents some protection from the cycle of financial panics that plagued the times. These gardens also gave each family a link with the land that was not available in the other town plats of Minnesota.

Another difference between this plat and plats for other new towns is the sheer size of the New Ulm plan. Most speculators tried to minimize their initial investment by platting small areas, practices that minimized the preparation cost and had the advantage of making the towns seem more successful than they actually were. A plat of two dozen blocks would fill up quickly, creating the need for additions to handle the burgeoning populations. Because speculators were not behind developing New Ulm, the city could be designed once for its likely maximum size. The town site was 40 blocks long and ten blocks deep, making it

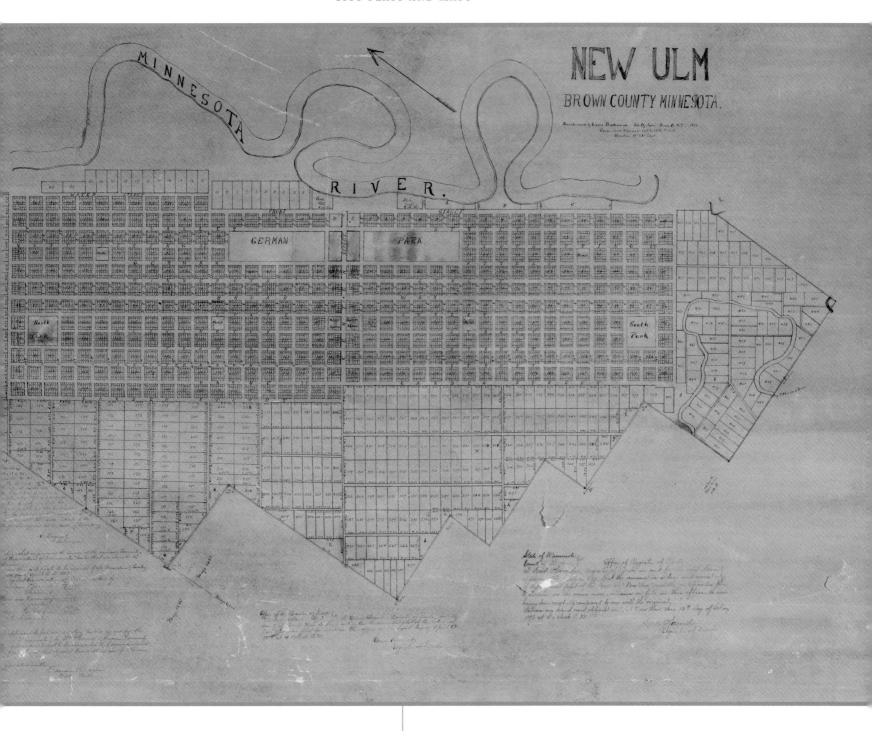

vastly larger than the plat for the territorial capital of St. Paul. The town designers may have created the gargantuan plat to keep the garden belt free from urban encroachment.

Other major differences may be seen within the actual layout of the city's 200 blocks and streets. First, instead of the main street running parallel and adjacent to the Minnesota River, which would be common for most river towns, the main street of Broadway was six blocks off the river on higher ground. Broadway was bisected by Center Street, which ran from the levee at First Street to the garden district. Although this crossroads would seem to be the most accessible point in the city, it was not selected for the main public space. North

German Park and South German Park, each six blocks long and two blocks deep, provided large community open spaces. The parks were separated from Center Street by a block intended to hold both the county courthouse and the school. Space was also set aside for the Turner Hall in this area. In addition, roughly each quarter of the city had a block set aside for markets. These four areas seem to have been intended to provide an alternative to the more typical Main Street commercial district. The plat is also distinctive because, like railroad towns, it does not conform to the regular grid of the Public Land Survey. This gives the edge of the town a unique sawtooth appearance.

H. G. Schapekahm. *Map of New Ulm, Brown County, Minnesota.*

NEW ULM: H. G. SCHAPEKAHM, 1875

24 X 38 INCHES

AUTHOR'S COLLECTION

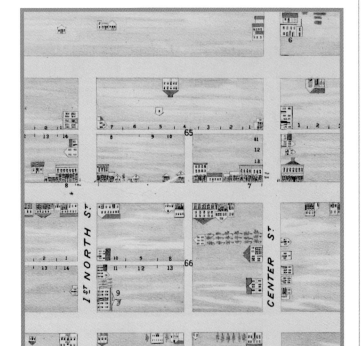

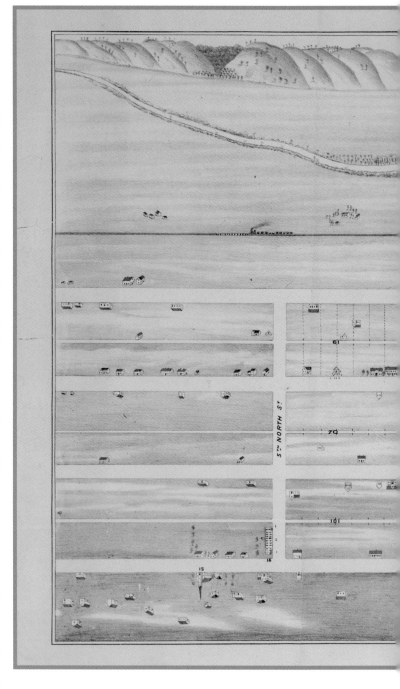

This distinctive picture map of New Ulm represents a rare effort by a local community to portray a snapshot of its existing development. In some ways, it resembles maps from insurance atlases of large cities, but it uses facade views instead of map symbols. The map contrasts sharply with the town maps in the contemporary Andreas *Atlas,* which are basically re-drawings of the official

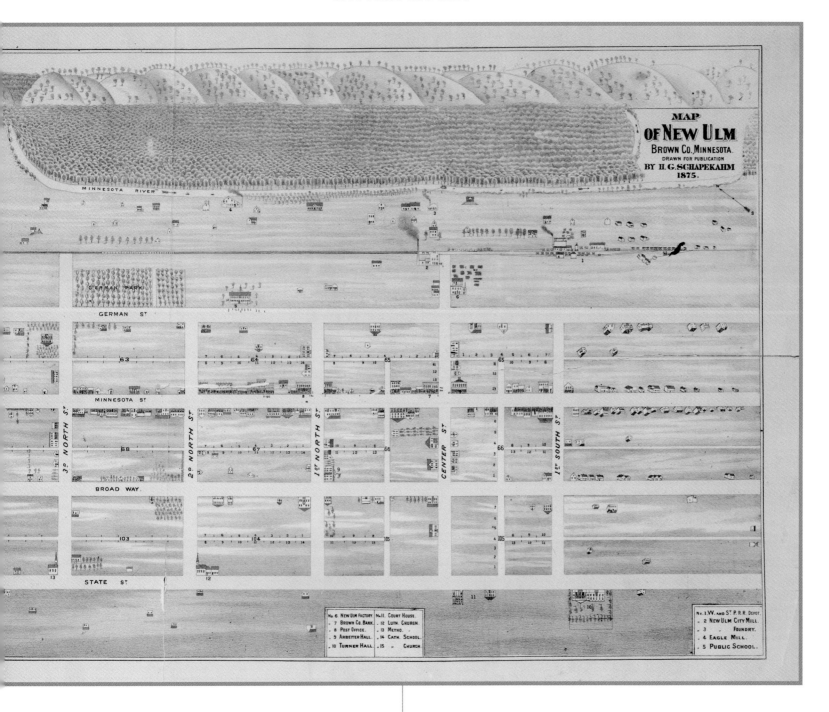

city plats. The map shows the sharp contrasts between the planning concept for New Ulm and the physical reality of New Ulm.

The motive for making this map is unknown, but it may have been drawn as an exercise in architectural drafting. It is the work of a young immigrant, Herman Schapekahm, who arrived in New Ulm in 1871 at age sixteen. In 1878, he went to St. Louis to study architecture, returning two years later to begin his career as an architect and contractor. He designed several landmarks in New Ulm, including Old Main at Dr. Martin Luther College and the houses of local brewers Otto and August Schell. He also was employed as the construction contractor on several other New Ulm buildings before his death in 1912 from a fall on a construction project.

H. Wellge. *Perspective Map of Duluth, Minnesota.*

DULUTH: DULUTH NEWS CO., 1887

18.5 X 41.5 INCHES

MINNESOTA HISTORICAL SOCIETY

Minnesota has one great port city, Duluth. When Duluth's port combined with the neighboring port in Superior, Wisconsin, more than a century ago, the Twin Ports of Lake Superior became one of the world's leading ports in terms of tonnage shipped.

Cities that host major ports have special geographic features and good connections to markets and areas of production. Leaders in port communities continually wrestle with problems of what goods to ship, how to improve facilities, and how to develop transportation systems to connect their port to larger trade areas.

The earliest European visitors to the site of present-day Duluth, which sits at Lake Superior's most westerly point, recognized its great potential. The first commodity they shipped was furs, provided by ancillary posts near the American Fur Company's post at Fond du Lac. Two dramatic landscape features dominate Duluth, the lake and the hill rising from it. The only flat lands available for growth are on Minnesota (Park) Point, Rice's Point, and the narrow west bank of the St. Louis River. The illustrator of this 1887 perspective rendering minimized the problem, perhaps because he wanted to show the city in a most favorable light.

Duluth's unusual topography was of little consequence during the fur trade era, but it posed formidable problems during the time of the development of the city and its industrial base in the nineteenth century. After the lakeshore Ojibwe ceded their lands to the American government in the mid-1850s, entrepreneurs eagerly established eleven separate town sites along the lakeshore. Each town was expected to boom because of the timber industry, the second commodity that provided a base for shipping out of Duluth. The future of

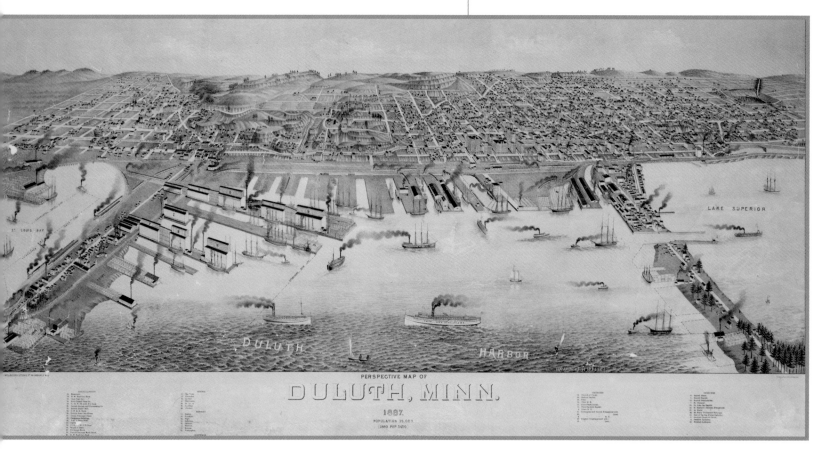

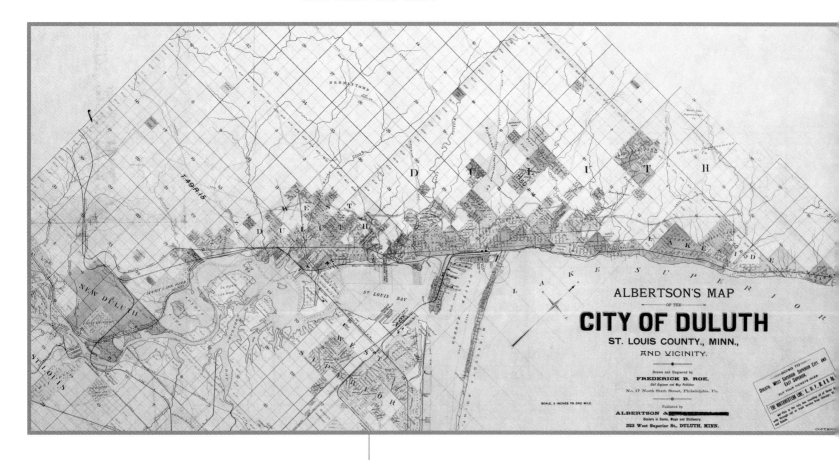

Frederick B. Roe. *Albertson's Map of the City of Duluth, St. Louis County, Minnesota, and Vicinity.*

DULUTH: ALBERTSON & CHAMBERLAIN, C. 1891

22 X 43.5 INCHES

MINNESOTA HISTORICAL SOCIETY

economic development seemed more enhanced by the building of a canal and set of locks at Sault Ste. Marie, on the eastern end of the lake, facilitating transportation between Lake Superior and Lake Huron. Unfortunately, the Panic of 1857 nearly wiped out the settlements at Duluth, and most new residents departed. Not until after the Civil War did the prosperity of the copper and iron mines in the Upper Peninsula of Michigan create a demand for a host of products, which many businessmen believed Duluth could provide cheaper than Chicago and other eastern cities.

In 1870, a railroad connecting Duluth to St. Paul gave the northern lake city its long-awaited link to the Mississippi River. It also enabled Duluth to take over some of the business of supplying the copper and iron ranges along the south shore of Lake Superior. In addition, the Northern Pacific Railroad, chartered in 1864 to become the first northern transcontinental railroad, finally began construction on its tracks west from Duluth. It would take another thirteen years to build the Northern Pacific far enough westward, through hundreds of miles without agricultural settlements, towns, or mines, to begin to create a flow of cargo for Duluth's port.

Construction of the Northern Pacific proved to be a mixed blessing for Duluth. It opened up a larger trade area, but the panic that resulted from backer Jay Cooke's financial woes just about destroyed the town. In fact, the City of Duluth surrendered its charter and shrank back to its village boundaries and political status. Ten years would pass before the rest of the town sites would be annexed into Duluth, known at the time as Zenith City. The 1891 Albertson map shows eleven communities that combined to become Duluth.

As the western railroads linked Duluth to the newly booming grain country of the Red River Valley, wheat took over as the third commodity to dominate the city's waterfront. Wellge's bird's-eye perspective map of 1887 and Albertson's map of 1891 clearly indicate the presence of the large grain elevators on the waterfront. The Bay Front District had multipurpose docks, special coal docks, and a dock owned by the railroad. Rice Point had three large grain elevators, two sawmills, and a blast furnace. Other sawmills dotted the shore toward Fond du Lac. In addition, elevator "A" sat outside the protection of the harbor. With the construction of these elevators, the dreams of the great port seemed possible, and the view shown in the Duluth and Superior map conveys the hustle and bustle of a booming harbor. The schooners that first opened trade on the lakes are still evident, but the presence of several steamships suggests that the sailing era is about to end. Dredges are at work creating new land, tugs pull the schooners out of the docks, and even a few intrepid recreational sailors ply the waters of the harbor.

Duluth's grain and timber trades continued in the late nineteenth century, but they were overshadowed by the iron ore trade. A blast furnace built by the Duluth Iron and Steel Company on Pike's Point in the 1860s seems to have been operational when both the Albertson and Wellge maps were published, but Duluth's fame would rest on shipping mountains of iron ore rather than on being a port that processed raw materials and shipped finished goods. And what an ore port it would become. The small but serviceable slips and docks of the timber and grain era would in time be overshadowed by the great wooden docks and trestles built in West Duluth to handle the specialized trains that came from the mines heavily laden with ore and returned empty. In 1893, the Duluth, Missabe, and Northern Railway built an ore dock 2,300 feet long, but even that would not be big enough, and eventually the wooden docks were replaced by steel facilities with a much larger capacity. Well over one billion tons have been transshipped from rail to freighter at the docks, and the tonnage grows every year.

The success of the port of Duluth was not solely the result of the deep and large harbor, which is protected from the storms of Lake Superior by the slim land spit at Park Point. Harbor improvements were needed, and Duluth's crafty political defiance played a major role in the port's success. After

Map of Duluth (Minnesota) and Superior (Wisconsin), Showing Lines of the Duluth Street Railway Company.

DULUTH: DULUTH STREET RAILWAY CO., 1917

17 X 15 INCHES

NORTHEAST MINNESOTA HISTORICAL CENTER, UNIVERSITY OF MINNESOTA, DULUTH

the Civil War, Congress ordered the Army Corps of Engineers to build piers at the natural harbor entry opposite the city of Superior and dredge channels along the 7-mile length of Superior Bay to Duluth. This configuration would have made Superior the dominant port city. Unhappy about this proposal, the Minnesota legislature in 1870 chartered the Minnesota Canal and Harbor Improvements Commission to dig a canal. Fearing for the future of the port at Superior, Wisconsin leaders filed a successful suit in state and federal courts to have Minnesota's dredging stopped. But Duluth was a long, long way from Washington, D.C. Although a telegram immediately conveyed the bad news to Duluth, it took the federal marshal three days to reach the city to enforce the order, and three days proved to be just enough time for Duluth locals, working nonstop with machines, picks, and shovels, to dig a channel near the Duluth end of the spit. The marshal arrived just a short time after the spit had been breached by the channel that would funnel ships directly toward Duluth's port. Because the harbor was slightly higher than the lake's level, water pressure quickly widened the channel, and the marshal was treated to the sight of celebrants watching the first ship, the steam tug *Fero,* pass through the new Duluth Ship Canal. Since 1872, the Twin Ports cities have shared the harbor with two entries but not always harmoniously.

The colorful Duluth street railway map from 1917 shows the two passages through Minnesota Point, one at each end, and the broader natural passageway into St. Louis Bay. It also captures the two cities' industrial landscapes at their largest expanses. Lumber, coal, and ore docks line the shore, and the Morgan Park steel and cement factories were beginning to produce their important products for the thriving towns.

CITY PLATS AND MAPS

Morgan Park, Duluth, Minnesota.
Dean & Dean Architects, Morrell & Nichols Landscape Architects, Owen Brainerd Consulting.

FROM *MORGAN PARK BULLETIN*, 1917

7.5 X 24 INCHES

NORTHEAST MINNESOTA HISTORICAL CENTER,
UNIVERSITY OF MINNESOTA, DULUTH

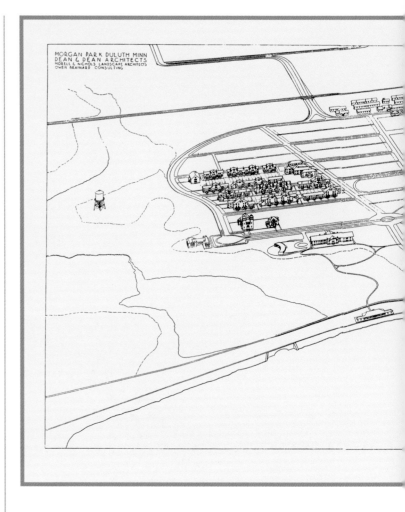

In 1907, United States Steel announced it would build a "monster plant" in Duluth for the manufacture of steel. One of the most important issues for the factory's success was securing a permanent labor force in an area where most people worked seasonally because of the severe climate. Interested parties believed that creating better housing conditions for workers than customarily available in company towns and other settlements could secure a stable all-year workforce for the area.

Morrell & Nichols, a Minneapolis landscape architecture firm, which had designed a large number of college campuses and had been involved in the development of a Minneapolis city plan, worked as consultants to various state agencies hired to do the Morgan Park plan. The Chicago architecture firm of Dean & Dean produced the design for Morgan Park's buildings. Design work began in 1913, and the simple housing was occupied beginning in 1914. Professionals at the time thought of Morgan Park, named for the U. S. Steel financier J. P. Morgan, as an example of what could happen when the best design principles were used. The broad streets, spacious lots, parks, and other elements of the infrastructure were thought capable of improving or maintaining the health of the workers. Provision of a school, library, and space for social clubs would promote the education and community life of the residents. The curved streets were intended to reflect the local topography and the lake. The extra-wide main street was the location for all the business and social spaces in the town. Houses were carefully designed to maximize space, light, and ventilation. All buildings were made from concrete and stucco, making the city very uniform—and somewhat drab. Dean & Dean planned to eliminate that problem by artful landscaping.

On this plan, company headquarters is slightly upstream. The steel mill itself is farther upstream from the town and beyond the border of this map. The cement factory was located between the railroad and state highway in the blank space on this view just upstream (to the left) of the main road leading into Morgan Park. In order to make efficient use of space, only two churches were built, the double-towered Church of the Blessed Saint Margaret Mary on the factory side of town and the United Protestant Church on the other side. Near the Protestant Church on large lots facing Spirit Lake were the houses of the plant managers. The company store and several independent enterprises in town were all housed in the same facility. Efficiency was also the watchword in education. The students attended in shifts during the day, and self-help adult programming was held in the school at night. As with most company towns, the sale of liquor within its boundaries was prohibited. In 1919, near its population peak, about 2,100 peo-

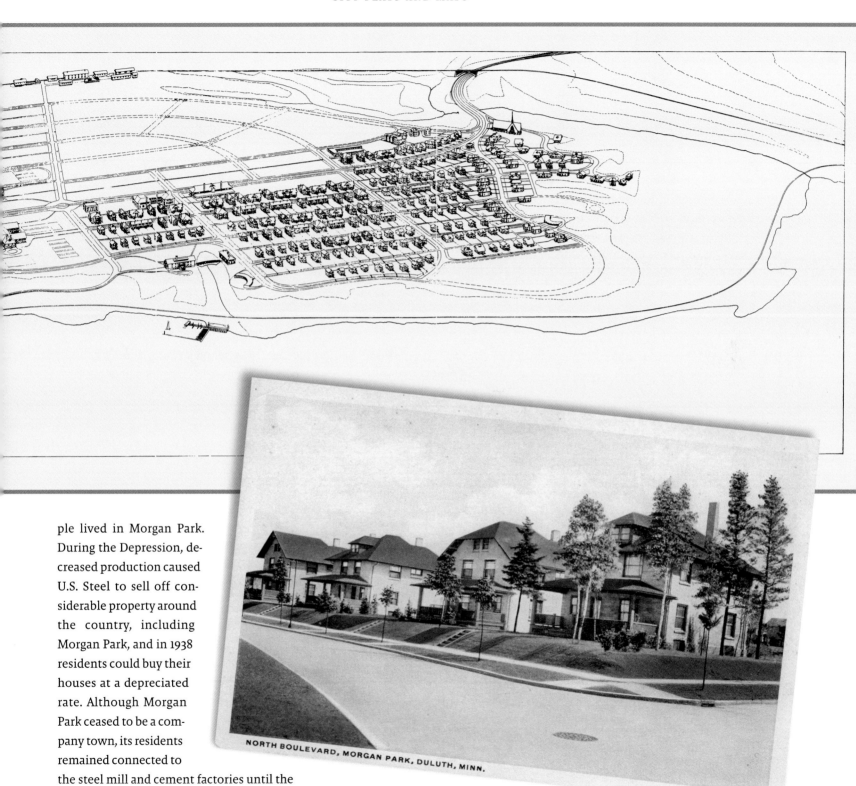

ple lived in Morgan Park. During the Depression, decreased production caused U.S. Steel to sell off considerable property around the country, including Morgan Park, and in 1938 residents could buy their houses at a depreciated rate. Although Morgan Park ceased to be a company town, its residents remained connected to the steel mill and cement factories until the 1970s, when all operations at the "monster plant" ceased.

Plat of the Town of Beaver, from *Plat Book of Winona County, Minnesota.*

MINNEAPOLIS: C. M. FOOT AND J. W. HENION, 1894

3 X 5 INCHES

MINNESOTA HISTORICAL SOCIETY

Current maps of southeastern Minnesota are peppered with place-names that commemorate towns that did not flourish—places like Black Hammer, Ostrander, Prosper, London, Easton, Alma City, Merton, Trosky, St. Killian, Bechyn, Swift Falls, Roscoe, and Skyburg. All were expected to grow and prosper, nurture children, and provide a final resting place for generations to come. But if left out of the emerging transportation system and new economy, they are recalled today only by a name on a map and a building or two, or perhaps some disintegrating ruins. Still others have a relic street pattern and are occupied by a few families who drive to work in larger towns nearby. The stories of the failed towns are similar: hoped-for railroad connections that did not materialize, competition from better-sited towns nearby that drew away the local market, changes in agriculture that created fewer demands for services, improved roads that led local populations to shop in larger towns and bypass local main streets.

The hamlet of Beaver was platted in 1856, and as the agricultural economy of the Whitewater Valley prospered, so did the town. By 1857, Beaver had a gristmill and a blacksmith shop. Farmers grew as much wheat as they could to supply towns like Winona and St. Charles with a burgeoning wheat trade. However, the engineers routed the railroad built west from Winona well away from this rugged topography.

After the soil originally deposited by the Whitewater River had been cleared and farmed, each new crop removed more and more of the land's fertility. As this happened, the soil also lost its ability to hold moisture, and with each rainstorm the water was able to flow more and more rapidly into the stream, in turn taking

Bird's-Eye View of Farm and Residence of H. B. Knowles, Whitewater Township, Winona County. The farm was a short distance down the Whitewater from the town of Beaver. Note the lack of forest in the scene.

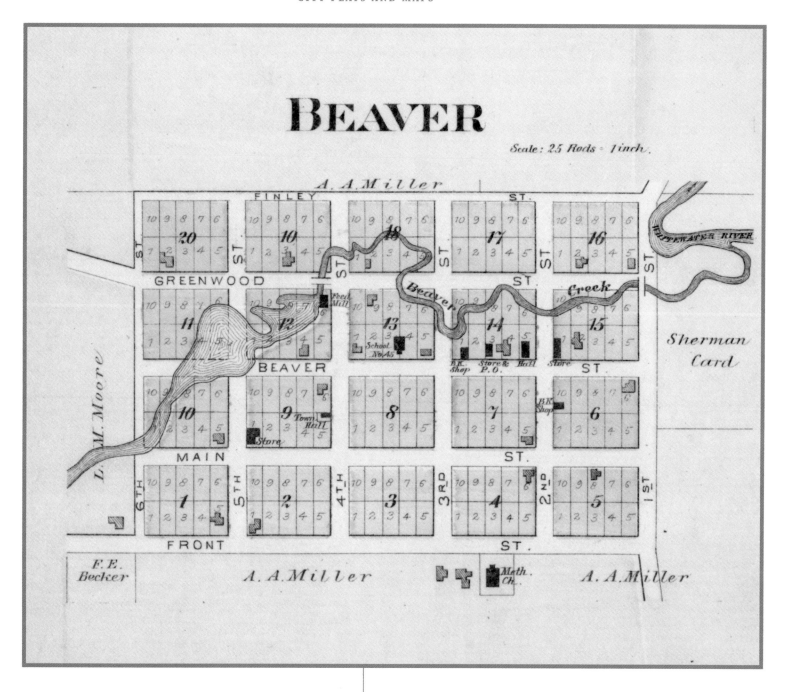

more and more of the soil. The ruining of the soil soon created a hopeless situation for the farmers of the Whitewater Valley. An initiative to preserve the landscape began in 1919 when Whitewater State Park was created; by 1927, it comprised 647 acres of land. Most of the valley was privately owned, and, in 1932, with the creation of a game refuge, the state began to buy up land as the farmers moved out. In 1938, the town of Beaver flooded twenty-eight times. This was the end of farming in the Whitewater Valley, and the towns of Beaver and Whitewater Falls were sold to the state. Today, the only structures standing are the Township Hall and a bridge over Beaver Creek. The old millpond is filled with silt, but the cemeteries remain to commemorate the human struggle to make the land productive.

Map of Wasioja Township, from *The County of Dodge, Minnesota.*

ST. PAUL: R. L. POLK & CO., 1905

16 X 12.5 INCHES

MINNESOTA HISTORICAL SOCIETY

In 1854, early settlers from New England and New York and other eastern states founded Wasioja, a town named according to the Dakota pronunciation of the Zumbro River. Once a prosperous town like nearby Mantorville, Wasioja was considered a fledgling city with a prosperous future to rival Rochester's. The business district housed law offices, a dozen shops, a hotel, a blacksmith shop, a gristmill, a furniture store, and a newspaper, the *Wasioja Gazette*.

Wesleyan Methodist Seminary, Wasioja, Dodge County. Forward-looking residents constructed substantial buildings in Wasioja, including the Wesleyan Methodist Seminary, which had, according to this view, a stupendous belfry. By 1874, the school was struggling, and it soon closed. The building burned in 1905.

In 1859, residents unsuccessfully attempted to relocate the Dodge County seat from Mantorville to Wasioja. The disgruntled losers accused the winners of fraud. Although Wasiojans mourned the lost county seat, their sorrows were forgotten among dreams of the new seminary and the railroad route that had been staked through town.

In 1858, the Free Will Baptists who had founded Wasioja began construction of the seminary with the goal of fostering religious educational opportunities. The large limestone structure opened in November 1860 with nearly 300 male students. By 1861, the school had been renamed Northwestern College and offered primary and collegiate education.

In 1861, Governor Alexander Ramsey was in Washington, D.C., when the Civil War broke out, and he immediately offered 1,000 soldiers to fight for the Union Army. In less than two weeks, Wasioja had the state's first recruiting offices. According to Wasioja legend, a professor from the Methodist seminary marched his entire class to the recruiting station for enlistment in the war. Nearly 200 men were sent from the town of Wasioja and other area communities.

By the time this plat map was made in 1905, the grid pattern of the streets in the town remained but all other evidence of a thriving community had evaporated. Wasioja had lost the battle to become the county seat, and the Civil War had had a devastating effect on the male population, nearly emptying the seminary and other economic institutions. The final blow was the lack of good transportation corridors to and from the city that would have permitted the transfer of goods and services to other markets. The Winona and St. Peter Railroad now traveled through Claremont, Dodge Center, Kasson, and Byron before entering Rochester but completely missed Wasioja. Nearby limestone quarries could not be exploited because of the distance from transportation opportunities. The history of the town, as told in the *History of Dodge County,* indicates the necessity of a well-placed bribe if small rural communities were to enjoy success and longevity on the rough-and-tumble frontier: "It is perhaps proper to state in this connection, that if the leading business man of the village at that time, Mr. Churchill [primary business owner] had seen fit to make a reasonable present of money or landed interests to the engineer, that the history of Wasioja might possibly have been different in some respects."

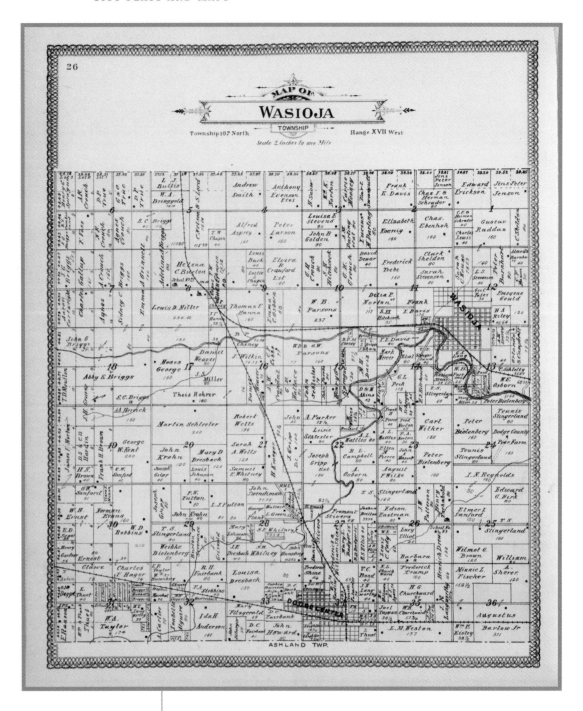

Unwilling to pay for votes in the election for the county seat or to bribe the railroaders, the leaders of the town were without either a connection to distant markets or the lucrative center of county government. Because of their scruples, Wasioja had little to hold it together when the Civil War recruited soldiers.

7 MAPPING THE TRANSPORTATION CONNECTIONS

Native Americans who once lived in what is now Minnesota developed a network of paths that connected important sacred sites, hunting grounds, and waterways. European and métis fur traders traveled along this same network, especially along the Rainy River route from Grand Portage on Lake Superior to Fort Gary on the Red River of the North. This Voyageurs Highway consisted of a string of lakes and rivers connected by portages or improved trails that enabled the men carrying heavy packs and canoes to move quickly overland. The system of trails and waterways was inadequate, however, for the area's new agricultural and urban settlers, who depended upon steamboats and wheeled vehicles drawn by draft animals.

In the 1830s, fur traders, who drove simple oxcarts along three routes to St. Paul from the Red River Colony north of Pembina, developed the earliest long-distance trails. The Hudson's Bay Company held the British monopoly on the fur trade in the lands whose waters flowed to the bay, but the company could not provide a strong enough market to absorb the products of the Red River Valley. Accordingly, in 1835, the first group of independent traders drove a caravan of oxcarts to the American Fur Company traders at Mendota. After Congress signed a treaty with the Dakota and the Ojibwe ceded their lands east of the Mississippi, in 1837, St. Paul became the region's dominant commercial center.

The main Red River Trail followed the line of the ancient beach of glacial Lake Agassiz, on the west side of the Red River. Upon reaching the headwaters of the Minnesota River, it proceeded downriver, connecting the American Fur Company trading posts in that region. The trail carried a large number of carts. It was usually dry, well marked, and easy to follow. The grasslands through which it passed provided plenty of food for the oxen. However, the trail passed through Dakota lands, and most of the traders were Ojibwe or métis with French and Ojibwe ancestry. In order to avoid the conflicts that developed, a second trail was established on the northern edge of Dakota Territory. This shorter and safer trail, opened in the 1840s, ran from present-day Breckenridge on the Red River to what is now St. Cloud on the Mississippi.

In 1844, the Dakota attacked caravans on this trail, stranding a group of traders in St. Paul. In order to return to their homes, this group opened a third trail, called the Woods Trail, through the northern forests of the Ojibwe. This route ascended the Mississippi to the trading post of Crow Wing, near Brainerd, and then turned west to what is now Detroit Lakes, where it continued northwestward to Pembina. This swampy, heavily wooded route was abandoned when a truce ended the conflicts between the Ojibwe and Dakota. These unimproved trails were vital to the economy of the new state. In 1857, at least 500 carts loaded with pelts and buffalo hides arrived in St. Paul. The development of scheduled packet-boat service on the Mississippi led the Hudson's Bay Company to abandon its policy of shipping all its goods out of British America via Hudson Bay, because the route to Europe was faster through the United States. The trails were the most important overland trade route in the Midwest and Canada.

The army built the first improved roads in Minnesota, connecting Fort Snelling to Lake Calhoun and the Falls of St. Anthony. A stage road established on the east side of the Mississippi in 1848 joined Galena, Illinois, to St. Paul via Prairie du Chien and Stillwater. It seems to have been used mostly in the winter. In 1838, the federal government created the Topographical Engineers Bureau and charged it with the task of conducting scientific reconnaissance and surveying wagon road routes that would expedite westward movement. When some concern was expressed about the legality of the federal government building roads within states and territories, the roads were labeled "military roads," although anyone could use them.

Minnesota had a very thin network of roads when it became a territory independent from Wisconsin in 1849. After the international border was established, a road was built in the far northeast from Fort William in Canada to Grand Portage. This road may have been more properly called a trail because most of the traffic was by foot. A road and a ferry linked St. Paul across the Mississippi to Fort Snelling. Another went from St. Paul to Stillwater; still another followed the old oxcart trail to St. Anthony and up the east bank of the Mississippi to Crow Wing, where it connected with the informal oxcart trail to Pembina. A road also extended from the St. Croix River east and south to the settled areas in Wisconsin.

As a territory, Minnesota represented a locale where the army could experiment with road survey and construction techniques and freely build roads under the guise of military defense. In 1850, Congress passed a Minnesota Road Act, containing an appropriation of $40,000 for a military road system. This system was to be an elaboration of the existing set of roads focused on St. Paul. Money was earmarked for the St. Croix Valley road from Point Douglas, just above the confluence of the St. Croix River with the Mississippi, to the St. Louis River. In addition, funds were available for work on the road from Point Douglas to Fort Gains (Fort Ripley), via Cottage Grove, Red Rock, St. Paul, and St. Anthony. A western branch of this road was to go from the fort to the Winnebago Indian Agency at Long Prairie. Funding was also authorized for a third road following along the west bank of the Mississippi from Mendota to Wabasha, and a fourth road from Mendota to the confluence of the Missouri and Big Sioux rivers at Sioux City, Iowa, was also included in the bill. The road from Point Douglas to the St. Louis River was thought to be especially important, because the timber companies needed a way to supply their logging camps during the winter.

In an attempt to expedite the building of roads in Minnesota, the 1851 Territorial Legislature required all healthy men from twenty-one to fifty years of age to work three days a year building roads. But the work was not completed fast enough for settlers, who clamored for them, nor did Congress ever appropriate enough money for the road-building effort. After making an initial appropriation, Congress failed to approve two subsequent funding requests. The construction of the roads was difficult, and not much could be accomplished during one construction season.

Map of the General Government Roads in the Territory of Minnesota, September 1854, accompanying the report of the Bureau of Topographical Engineers, Senate Executive Documents, no. 1, 2nd sess., 33rd Congress.

19.5 X 13.25 INCHES

AUTHOR'S COLLECTION

The sluggish progress of government road building annoyed everyone in the territory. Frustrated over the War Department's unwillingness to sponsor more military roads in Minnesota, the territory's delegate to Congress, Henry M. Rice, pushed the Department of Indian Affairs to build the roads that had been promised to the Ojibwe as part of their treaties. As a result, roads were built from Crow Wing to Leech Lake and from the confluence of the Mississippi and Rum rivers to the village on the west shore of Mille Lacs. Although the federal government appropriated more funds to road building in Minnesota Territory than to either Wisconsin or Iowa Territory, the proposed network was unfinished when statehood was achieved in 1858. With statehood, the national government's responsibility for the roads ended, and roads became the responsibility of the communities through which they ran. In 1861, the U.S. Army Road Office closed and all the property it owned was liquidated.

Because military roads were meant to be long-distance links between forts or settlements, they followed the rivers and did not necessarily provide good access to the best agricultural lands. As a result, counties, townships, and farmers laid out their own roads. For example, squatters in the Minnesota River Valley hired William Dodd and Auguste Larpenteur, in late spring 1852, to organize a road-building effort that cut a rough road 65 miles long between St. Peter and St. Paul. This road curved through the hardwood groves of beautiful rolling countryside, following the divide separating the Minnesota River from the Cannon and Vermillion rivers. It was built before the legislature authorized the road, and in fact most roads were authorized only after they appeared on the landscape. The civilian roads were necessary because the growing tide of immigrants could not move their livestock and farm machinery by steamboat. Unable to wait for perfect roads, the farmers moved west on the system of trails, paths, and lanes freshly cut through the big woods and prairies.

The growing settlements in south central Minnesota required better mail service and regular connections to the rest of the country. Therefore, the staging industry and public roads expanded dramatically after the Treaty of Traverse des Sioux in 1851 opened the southeastern portion of the state west of the Mississippi River and south of the Minnesota River. Roads were approved that connected the Mississippi River with the Minnesota River, passing through what are now Rochester and Owatonna; another road along the Minnesota River connected Fort Ridgley to the Twin Cities area.

The legislature authorized dozens of roads but did not appropriate funds for their construction. Therefore, the staging companies or county and township governments cleared and maintained them. Because the roads were so bad, stagecoach travel was not always feasible, and the passengers were forced to ride in "unsprung," or springless, wagons. In winter, travel was usually by sleigh. Eventually stage lines carrying mail as well as passengers reached the Red River Valley and connected most parts of southeastern Minnesota.

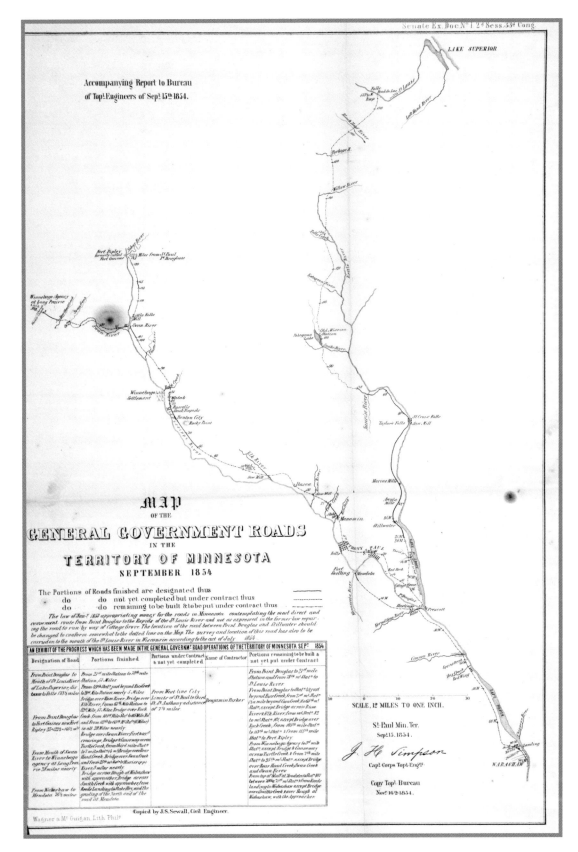

Map of Minnesota Territory, from
John Wesley Bond, *Minnesota and Its Resources.*

NEW YORK: REDFIELD, 1853

12.25 X 15 INCHES

MINNESOTA HISTORICAL SOCIETY

The influential book by J. W. Bond which contained this map, *Minnesota and Its Resources,* extols the qualities and potential of the new Minnesota Territory. Aside from the mosquitoes, the swamps, and the layout of the original plat for the city of St. Paul, Bond found little wrong with Minnesota—except for the roads.

Bond's narrative is based on his journal of travel from St. Paul to Pembina to witness Governor Ramsey negotiate a treaty with the Native Americans occupying the Red River Valley. The writer's account of the main road north, the Red River Trail, is a witty report on all sorts of tribulations and pleasures of a camping trip lasting several weeks. The trail was well marked but consisted of bogs, informal corduroy roads (logs laid side by side transversely to make a road surface), rudi-

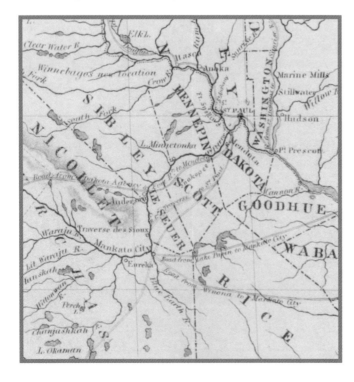

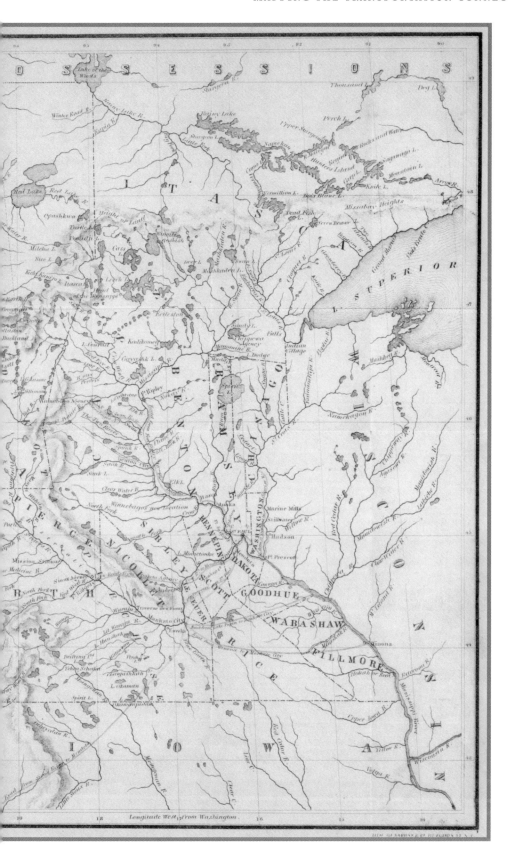

mentary bridges, dangerous fords, and the usual problems of mud and dust. Thus, he clearly knew that the roads shown on this map were more dreams and approximations than actual broad and improved roads leading from the river towns into the productive interior.

Although Bond's chapters on transportation focus on dreams of north-south and east-west transcontinental roads crossing in the Twin Cities, he devotes a few pages to the slowly expanding road system. In particular, he seemed excited about the progress being made on the road from Point Douglas to the St. Louis River: "Twenty-four miles of the Point Douglas and St. Louis River road have been constructed opening from Stillwater northward for that distance a good highway. The extension of this road is required to bring into market the extensive and richly-wooded, but inaccessible region, laying north of the Marine mills, and open to settlement and the enterprise of our lumbermen, tracts of valuable land now lying waste for want of means of communication with them" (Bond, 224).

In southern Minnesota, roads appearing on this map focus on connecting the river ports on the Mississippi to the rich land between the Minnesota and the Mississippi and on to Mankato and Traverse des Sioux. It is noteworthy that this map shows many more roads than do maps published in 1855 and later by the major atlas publishers.

Not surprisingly, the horse-drawn staging industry stagnated during the Civil War and the Dakota conflict. After the war, there were about 1,200 miles of stage roads in the state, but it was clear that the railroad would offer much better long-distance transport. In 1862, one 10-mile-long railroad in the state connected St. Paul and St. Anthony. By 1865, the network had expanded to 210 miles. By 1872, however, there were 2,000 miles of rails, and by 1880, 3,000. By the end of the century, the railroad system had doubled to over 6,000 miles of track. The roads built by the staging companies became feeders for the railroad and were of local interest only. ✸

Isaac Ingalls Stevens. *Preliminary Sketch of the Northern Pacific Rail Road: Exploration and Survey, Map 1, "From St. Paul to Riviere des Lacs."*

PHILADELPHIA: WAGNER AND MCGUIGAN LITH., 1853–1854

26 X 37 INCHES

MINNESOTA HISTORICAL SOCIETY

During the mid-1850s, national public opinion shifted from favoring expanding the size of the United States to favoring tightly binding the new territories to the more established states in the East. After railroads reached the Mississippi River in 1854, business interests began pressuring Congress for a plan and subsidies for expanding the railroad network into the trans-Mississippi West.

At first a plan was formulated to build three railroads, but Congress deemed that too expensive and decided that only one could be supported. This prompted a heated dispute between the county's free and slaveholding states. The dispute over the proposed railroad line was both local and regional, with numerous western and eastern terminal cities vying to be on the route. Congress was stalemated. Because some action was needed, the task of selecting the route was passed on to the War Department, and the engineers were ordered to pick the best route.

Secretary of War Jefferson Davis was charged with assigning survey teams to all prospective routes and selecting the best one based on data compiled by the field parties. To accomplish this huge task, Davis established the Office of Pacific Railroad Explorations and Surveys, which was directed by army engineering officers. Each expedition was required to report on the numerous factors affecting railroad construction. Essentially all details of the physical landscape had to be recorded, including distances, slopes, valleys, ravines, passes, canyons, potential bridge sites, and tunnels. In addition, each survey had to evaluate the economic potential of the route. Congress wanted to know about supplies of timber, stone, coal, and water. Large teams of scientists went along with the survey parties to carry out detailed observations. Thus, the railway surveys provided the first great opportunity for geography and other sciences to influence national policy.

Four potential transcontinental routes that had significant support among members of Congress were selected for surveys. The Minnesota-based route, between the forty-seventh and forty-ninth parallels, from St. Paul, Minnesota, to Puget Sound, was surveyed by former engineering officer Isaac I. Stevens, who was also the new governor of Washington Territory. In the spring of 1853, Stevens led his expedition, the largest and best equipped, west from St. Paul. Highly competent and an effective leader, Stevens had 240 men in his party—11 officers, 76 soldiers, and an assortment of scientists, teamsters, guides, and wranglers. He needed a force of this size because his was the most difficult route. For 40 years, no surveyors had ventured into this territory—still very much controlled by the Native Americans. Furthermore, the expedition needed to get through the high mountains before the snows of winter closed the passes.

Stevens divided his men into several detachments, each with its own assignment. While he went west across the plains, another group started from Puget Sound, and two others worked in the Rockies. Stevens wanted to have the northern route selected because it would ensure the success of Washington Territory. He casually overlooked the issues of high mountains and cold northern winters and focused on the broad plains, passes in the Rockies, and the connection with China. Some members of the survey team agreed with him; others thought the route impossible. The subsequent construction of the Great Northern and Northern Pacific

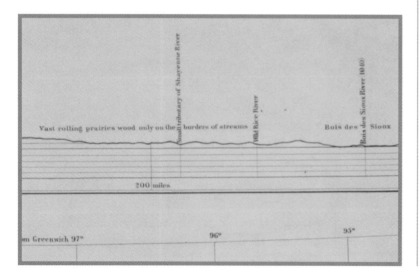

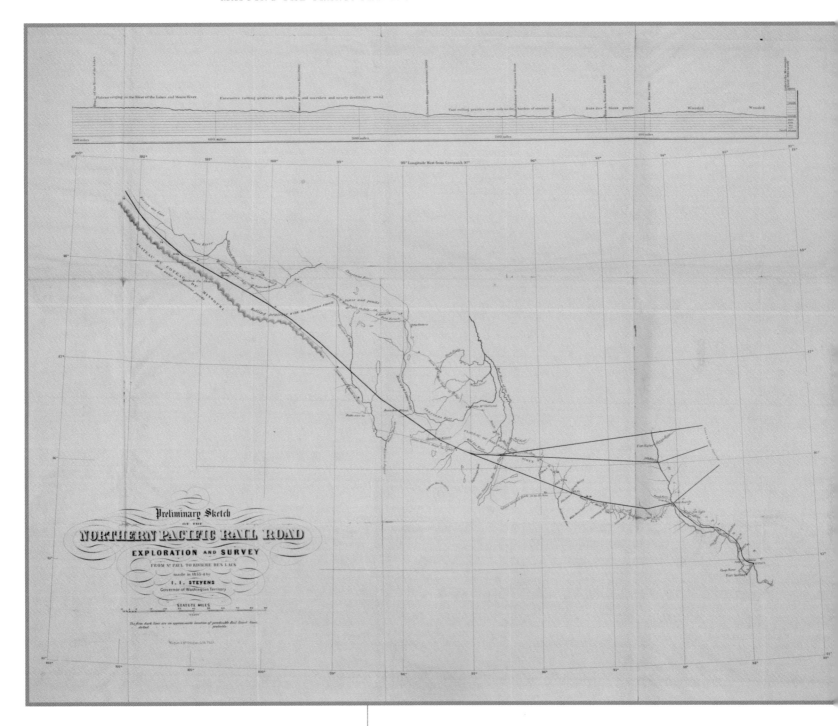

would prove Stevens correct, but those construction projects would demand leadership, creativity, and government financial support.

The three-sheet published map of the Stevens expedition survey shows the route and important topographical features along the way. This map is a very generalized compilation of the surveys, but it is detailed enough to trace the route on modern maps. Map compilers and publishers in the nineteenth century included this proposed route on their maps for decades.

Minnesota,

from *An Atlas of the Northwest.*

CHICAGO: RAND MCNALLY, 1896

18.9 X 25.75 INCHES

MINNESOTA HISTORICAL SOCIETY

Rand McNally & Company, a major name in map publishing to this day, began as a small shop opened in Chicago in 1856 by William H. Rand. The shop originally printed guidebooks and directories. In 1858, Rand hired Andrew McNally, who later became a partner. Within the next few years, the shop entered into the railroad business by printing railway tickets and schedules, which led to the publishing of railway guides, and by 1872, Rand McNally and Company had the capacity to publish maps. While continuing to produce railroad maps, pocket maps, and booklets, the company became a serious atlas publisher in the 1880s and added lithography to its printing capabilities.

What made Rand McNally unique at the time and gave it an important advantage in the map publishing business was the use of the wax engraving method called "relief line engraving." With this process, plates could be created very quickly and changed easily, necessary attributes during the time of rapid railroad growth and westward expansion. The wax engraving method also allowed the use of very small lettering, useful when many stops and stations along a route needed to be marked. This technique was not popular in Europe, where the smaller, more discerning market preferred the more elegant products of engraved maps. Map publishers in the United States, however, served a much larger mass market that was less interested in the refinements of the cartographic craft than in the locations of places. The great popularity of these maps so conditioned the consumers that other forms of cartography tried to imitate the wax engraving look and style. These maps could be filled with place-names, but the engraving technique did not show shading well, and thus topographic features were downplayed in favor of more and more names. Rand McNally & Company quickly realized the potential in the map market, and the company expanded rapidly.

This map illustrates the capability of Rand McNally to produce a variety of special-purpose maps for railroad clients and passengers. The atlas was intended to increase travel on the sponsoring railroads and was distributed to major shippers. The map also captures the networks at a critical time. The colors (purple, Burlington; green, Great Northern; and red, Northern Pacific) clearly identify the railroads that would be brought under the control of the Northern Security Company in 1904 by James J. Hill, J. P. Morgan, J. D. Harriman, and J. D. Rockefeller. These big industrialists wanted to gain a monopoly over transit in this region, even though the Sherman Antitrust Act of 1890 had outlawed this kind of practice because of its stranglehold on users.

Trust-busting president Theodore Roosevelt successfully sued Northern Security Company for violating the antitrust

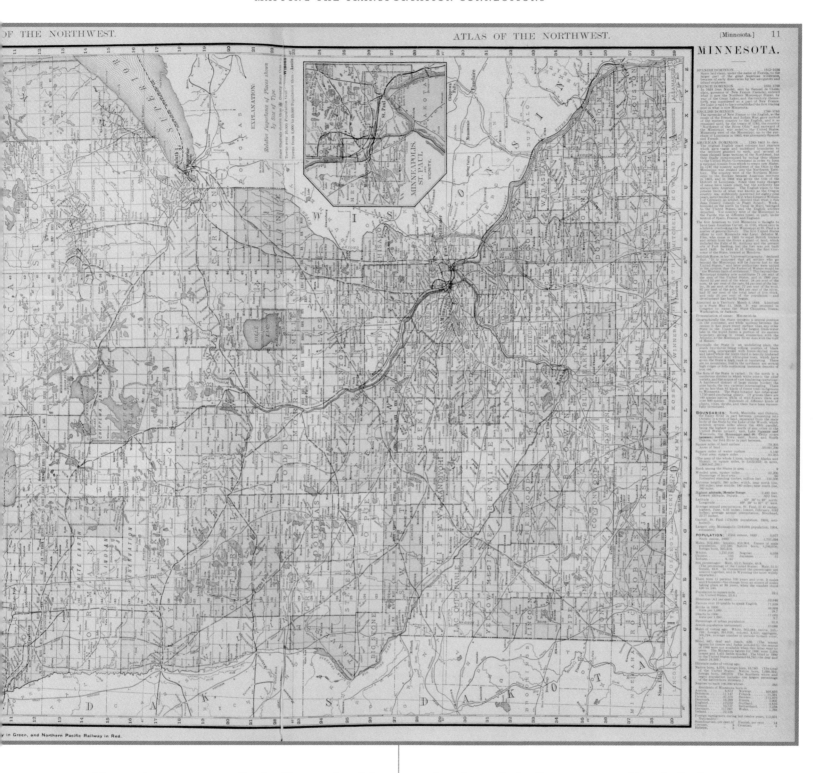

act, and in 1904 the company was dissolved by an order of the U.S. Supreme Court. The other railroad companies prospered for a few decades, but by the end of the century these railroads and several other lines were merged into the Burlington Northern and Santa Fe Line, and the miles of tracks were significantly reduced.

Auto Trails Map of Minnesota and Western Wisconsin, from *Commercial Atlas of America.*

CHICAGO: RAND MCNALLY, 1920

28.25 X 20.75 INCHES

MINNESOTA HISTORICAL SOCIETY

Rand McNally understood the need for new maps to complement its extensive railroad map publishing business. It accordingly began publishing a series of highway maps titled Auto Trails. In 1904, the company published a road map of New York to launch the new venture.

By 1920 a new map, showing the auto trails of Minnesota and western Wisconsin, marks the transition to the new concept of state and national roads. It shows the new Wisconsin road-numbering system, as well as the older trails and the main roads. Close examination reveals the frequent right-angle turns that were created when roads followed the township lines through the countryside. Travel was slow, but nevertheless the corners were a hindrance to speed. The map makes no qualitative distinction among the roads, so motorists were frequently surprised by the surfaces they encountered.

In 1920, there were very few drivable roads in the big bog area in the northern part of Minnesota. The North Shore highway headed inland to avoid the steep cliffs north of Two Harbors. In the southern section, the road net was denser, but a large number of roads appeared to come to a dead end when the improved surfaces ended. It is interesting to note that Rochester still is more connected to the river cities to the east and west than north to the Twin Cities. The presence of the railroads on the map clearly indicates their continuing importance to the state's transportation industry.

Douglas Lodge, Itasca State Park, about 1925

MAPPING THE TRANSPORTATION CONNECTIONS

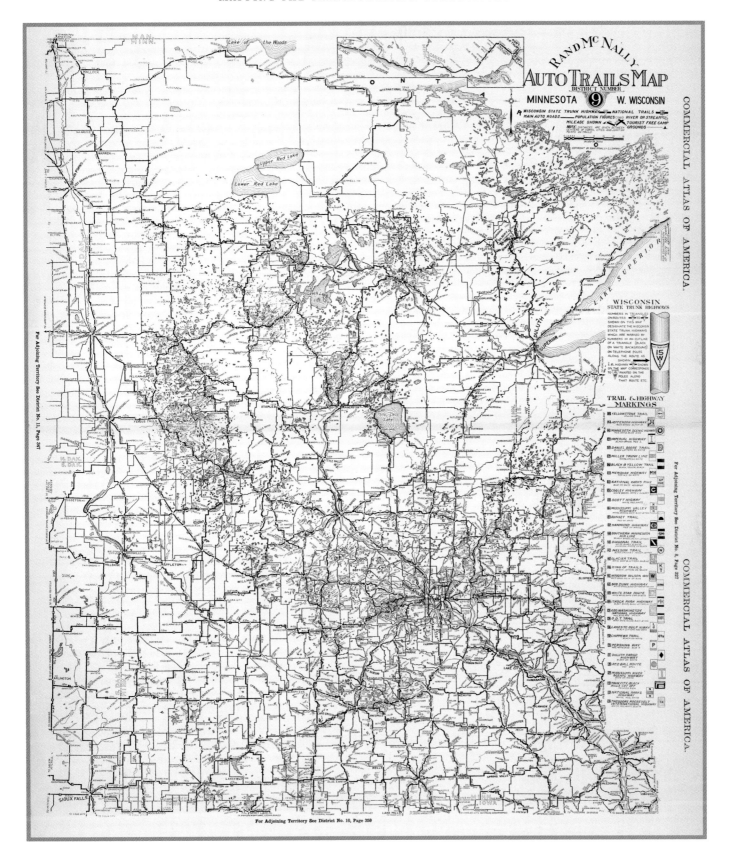

Early privately published road maps of Minnesota.

CRAIG SOLOMONSON COLLECTION

Throughout the nineteenth century, Minnesota's system of wagon and buggy roads was largely the responsibility of local governments. Anyone wishing to travel or ship goods for a long distance opted for the railroad. The quality of the local roads varied tremendously, but most were in poor condition. In wealthy communities, roads were graded and graveled. Other communities provided the best roads they could afford.

Outside large cities, there were no paved roads. Therefore, the best time to travel was during the winter, when the roads froze or were covered with ice and snow that made travel by sleigh or cutter efficient and fast. In the spring, roads developed large potholes, and once the roads dried out, they became sand traps, where horses' hooves and wagon wheels kicked up clouds of dust. Local farmers, as supervisors of the town roads in the township road districts, laid out roads along township lines and along the edges of farms. The resulting system was inconsistent and not well connected. Counties also took on road building and maintenance, but the results were similar. Poor roads, an inconsistent network of roads, and the lack of maps or guides hampered wagon and buggy travel.

Directions were all based on local landmarks and rough estimates of distance. A town might be a "fur piece" or "over yonder." Directions were related to barns, schoolhouses, forks in roads, and other landscape features. This was not a major problem for most people, because travel was largely local. In 1900, there were fewer than 75 miles of paved roads in the

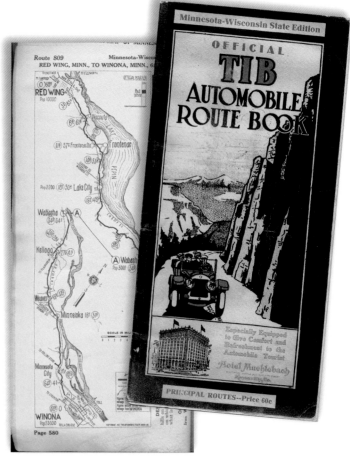

state, so people and products traveling long distances usually went by train.

But this changed when use of the automobile became more widespread after 1920. Into the map gap moved private enterprise and the automobile clubs, which began publishing directions and then combinations of directions and simple strip maps as itinerary maps. These groups also developed a system of named and marked roads that facilitated long-distance travel. The development of named trails began in the second decade of the twentieth century in response to the rapid increase in the number of car owners. The recreational auto era had begun.

Enterprising groups developed to expedite travel along certain routes and to lobby for improved roads. Nationally, the first two organized routes were the National Old Trails Road, running from Baltimore to Los Angeles, and the Lincoln Highway, at first called the Coast to Coast Rock Highway because the surface was gravel, which ran from New York to San Francisco. The Lincoln Highway opened in 1915 in time for the Panama-Pacific Exposition in San Francisco. The backers of the various routes selected a series of roads, christened their routes with memorable names, formed trail associations, and solicited businesses and towns along the way for dues. The associations promoted their routes with guides, newsletters, and conventions and encouraged the governments to begin to improve the roads along their routes. By the mid-1920s, there were 250 identified routes, running from the Atlantic to the Pacific and from the northern border to southern cities. One of them, the Jefferson Highway, passed through Minnesota on its way from Winnipeg to New Orleans. In Minnesota, there were many shorter routes, as well as regional trails, including the Red Ball Route (Cedar Rapids, Iowa, to St. Paul) and the Wonderland Trail (Duluth to Grand Forks, North Dakota). All were established to promote travel,

good roads, and the economic prosperity of the towns they passed through.

Signage was a huge problem, which each trail association tried to solve by painting its unique logo on any reasonably permanent landscape feature visible from the road. Telephone poles were the preferred spots for painting a sign, but anything—trees, barns, rocks, or bridges—would do when the poles were not available.

These trails provided a valuable service in the early days, but there were serious problems with the system. Some did not really meet a demonstrated traffic need; others went through as many dues-paying towns as possible rather than taking the shortest route. In other cases, corporations such as hotels distributed maps that were intended to bring travelers to their doors. Other retailers also distributed maps, but the connection between their products and travel is not so

obvious to us today. Rival routes frequently overlapped, making motorists uncertain as to which to take when the routes separated. Some wondered if the organizers were really providing a public service or just trying to make easy money by exploiting auto tourists.

The first state to replace the trail signs with numbers was Wisconsin, which created a state trunk highway system in 1917 that required uniform signage. In the Federal Aid to Highways Acts of 1910 and 1921, limited federal funding was provided to support the development of roads for the transport of mail, which previously had been carried on the trains, and for interstate travel. The 1921 act provided funding for 7 percent of each state's road network. The roads eligible for federal aid on a fifty-fifty basis were identified in 1923, and the long-awaited era of road building dawned. Long-distance travel, however, remained an adventure for many years to come.

State Highway Department Map of Minnesota Showing the Status of Improvement of State Roads, January 1919.

ST. PAUL: MINNESOTA HIGHWAY DEPARTMENT, 1919

19 X 15 INCHES

MINNESOTA HISTORICAL SOCIETY

This state-issued map is one of the most important Minnesota maps ever printed. It played a fundamental role in the debate about the state's future because it showed wary voters how the entire state could be linked by a system of roads, if they agreed to pass new legislation. The map demonstrated the scope of the proposed system to the public, which lengthy descriptions in words were unable to do. Three colors of lines showed the status of roads that travelers would encounter: partially constructed, graded, or graded and surfaced.

Building and maintaining a road system challenged all levels of government. To cope with the burgeoning requests for more and better roads, Minnesota created the State Department of Highways in 1917 under the dynamic leadership of Charles M. Babcock. He and his supporters in the Good Roads movement had been able to convince the legislature and the general public to change the state's constitution to establish a trunk highway system, consisting of 70 highways and approximately 7,000 miles. The arrangement transferred taxes from the counties to the state, which then assumed responsibility for state highways. Babcock was not finished, however, and eventually he got the state to reserve all its gasoline tax revenue for building and maintaining roads. In order to develop public and official support for what became known as the Babcock Plan, this full-page map of the proposed state road network appeared in newspapers, along with a concise statement of the important points of the legislation.

After Babcock's plan was approved, the state published an official state highway map, and each year thereafter a new map showed recent progress in extending the system. Larger cities were connected to one another first, and year after year the network improved. Only the dedicated gasoline taxes made it possible.

For a good many years, highway travelers commonly ex-

Since 1934, the covers of official state highway maps have been used to send powerful messages to travelers and residents. The maps themselves provide exact spatial information, but the covers invariably depict an attractive and interesting scene in good weather. Some covers highlight special events or eras in the history of the state; others highlight special landscape features.

perienced car troubles. Tires, in particular, frequently failed on long trips. Cars were relatively light, and when the paved roads ended, travelers took to the fields and farm lanes to reach their destinations. Visionaries predicted to their children that someday all roads would be graveled. Car camping became very popular, even for honeymooners, and maps began to show camps catering to travelers.

The success of the Federal Aid to Highways Acts of 1910 and 1921 and the rapid increase in the number of households traveling by car created the need for up-to-date highway maps, as well as an opportunity for gasoline companies to advertise. Petroleum companies contracted with map publishers to produce easy-to-read, efficient maps of national, state, and county highways in states or clusters of states. As might be expected, Chicago map publishers, facing waning demand for railroad maps, moved into this market. Major oil companies—Amoco, Standard Oil, Pure Oil, Sinclair, Phillips Petroleum, and others—franchised gas stations to sell their products.

In most service stations, a rack holding free maps hung by the cash register. The maps distributed by these different companies varied little, and their covers conveyed a variety of positive messages about driving, including happy and confi-

dent women driving their autos and happy families motoring through the countryside on vacation. Hundreds of thousands of these maps were produced for motorists, who wore them out or replaced them with newer editions.

In 1944, the Federal Aid to Highways Act provided much-needed funding for highway construction and improvement programs in Minnesota that had been delayed by the Depression and World War II. In 1956, the Federal Aid Highway Act, more popularly known as the National Interstate and Defense Highways Act, added another layer of superroads in the highway system. Although President Dwight Eisenhower and others intended this system to improve the movement of traffic across the country and between major cities, freeways obviously dramatically affected the cities themselves as well.

All these changes have been recorded on the official state highway maps produced by the Department of Transportation, which publishes maps every two years. These maps have become a primary record of place-names within the state. Many towns platted in the land rush of the nineteenth century remain only as names on the official map. Freeways are the dominant theme of contemporary maps, just as railways and state highways dominated the older maps.

MAPPING THE
DEVELOPING TWIN CITIES

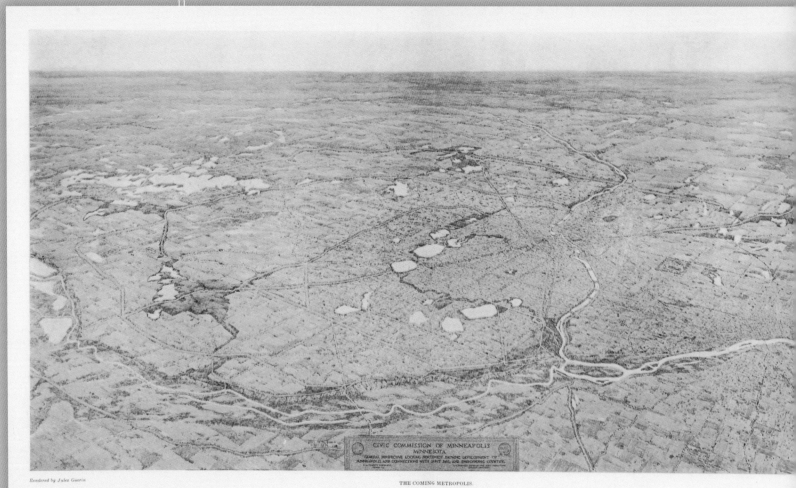

Rendered by Jules Guerin

THE COMING METROPOLIS.

Maps offer us the opportunity to visualize the many ways that people experience their environments. Over the past century and a half, a wealth of maps depicting the developing Twin Cities have offered visual assessments that range from reasonably accurate to wildly grandiose. Many of these maps document the cities' growth and how planners have attempted to manage it.

Centuries ago, Native Americans inhabited sites near the confluence of the Mississippi and Minnesota rivers. Minnesota's earliest maps of development date to 1805, when Lieutenant Zebulon Pike, acting on behalf of the U.S. Army, acquired two sites from Dakota residents of the land to be used as plots for future forts and their dependent settlements. One was at the mouth of the Minnesota River; the other was at the mouth of the St. Croix River. The map of the military reservation qualifies as Minnesota's first "urban" map. Once the acquired land became open to outside settlement, parcels sold relatively quickly, and before long developers were subdividing farmlands around the Twin Cities into residential communities. The resulting plats represent the second phase of urban development maps. Some of the plans were highly exaggerated for their time and place. The story of these dreams and schemes is told in the re-platting of communities.

One of the best ways to understand the growth of cities is to view them as responses to developments in transportation and communication technologies. Using this concept, it is possible to divide the history of American cities into three phases: steamboat and walking cities; streetcar and railroad cities; and auto and freeway cities.

During the first years of the territory and state, long-distance travel in Minnesota depended on steamboats, and local travel depended on humans and animals walking. Accordingly, cities were small and crowded close to their waterfronts. Gradually railroads replaced steamboats for long-distance shipping and travel. Commuters used trains to some degree, but it was electric-powered streetcars that ushered in a new concept of city life and prompted all sorts of new land uses. These changes and the expansion of steam-powered industry resulted in several attempts to plan and control the growth and quality of the cities.

The most recent great change in the Twin Cities urbanized area came with widespread use of the automobile and the building of interstate freeways. These alterations created a period when the cities expanded rapidly in all directions. Changes at the edges of the cities also resulted in changes at their cores. The downtowns surrendered most of their commercial functions to suburban shopping centers, and city dwellers seemed to consider older residential areas less attractive than new suburban subdivisions. These changes promoted another era of planning and the formation of the Metropolitan Council. The creative solutions for many urban issues developed by the Met Council and the state's political leaders have made the Twin Cities a model of urban planning.

The Twin Cities development maps included in this chapter portray landscapes and environments ranging from immigrant squatter communities to the most elite neighborhoods, from the commercial core to districts of heavy industry. Some show only a small section of the Twin Cities; others cover the entire metropolitan region. They suggest some of the many ways to experience and envision an urban center.

Jules Guerin.
The Coming Metropolis: General Perspective Looking Northwest Showing the Development of Minneapolis and Connections with Saint Paul and Surrounding Country,
from Edward H. Bennett and Andrew Crawford, *Plan of Minneapolis.*

MINNEAPOLIS: CIVIC COMMISSION, 1917

12.25 X 18.5 INCHES

AUTHOR'S COLLECTION

Lieutenant J. L. Thompson.
Map of the Military Reserve Embracing Fort Snelling.

1839; COPIED BY WILLIAM GORDON IN 1853

21 X 21 INCHES

HENNEPIN COUNTY SURVEYOR'S OFFICE

Thirty-four years after Lieutenant Zebulon Pike purchased a vaguely bounded area at the confluence of the Mississippi and Minnesota rivers, a limestone military fortress named Fort Snelling dominated the surrounding river valleys and protected the interests of American fur traders. The military reserve was much larger than the actual fort because the solders needed wooded land to harvest for constructing and heating buildings. They needed gardens to supplement their meager army rations on the frontier. Civilians lived on the military reservation, as squatters who provided a variety of goods and services to the soldiers. Across the Minnesota (St. Peter's) River and squatting within the military reserve boundaries was a fur trading post with solid limestone buildings established at Mendota by Henry Sibley.

The civilian population near Fort Snelling began to balloon when disillusioned members of the Canadian Selkirk Colony at Fort Gary migrated south into the vicinity and down the Mississippi. Just how many arrived is unknown, but by 1827 there were several farms on the government land. Most were still in operation ten years later when the Ojibwe Treaty of 1837 opened land east of the Mississippi to settlement.

When Major Joseph Plympton arrived to take command of Fort Snelling that year, he set about determining appropriate limits for a military reservation and for civilian settlements. The military reserve area was cleared of Dakota title and opened to settlement, but some of it was reserved from the land-sale process and held for military purposes. Plympton had Lieutenant E. K. Smith conduct a survey and then forwarded the results to the War Department for action. In his accompanying letter, he expressed concern that there was not enough timber in the land that he thought was to be included in the area referred to in Pike's Treaty. The War Department ordered him to determine the amount of land needed.

Lieutenant J. L. Thompson conducted another survey in October and November 1839. His work provided the basis for this *Map of the Military Reserve, Embracing Fort Snelling.* According to Major Plympton, the boundaries shown on this survey slightly expanded Smith's assessment to include certain existing but unwanted settlements, including the drinking establishment of whiskey trader Pig's Eye Parrant at Fountain Cave. Despite formal protests by squatters to officials in Wisconsin Territory, the reserve as marked out by Thompson was established. A detachment of soldiers from Fort Snelling forcibly cleared settlers from the reserve, but the squatters did not move far. They reestablished themselves in a settlement on land outside the reserve in what is now downtown St. Paul. Thus began the legend (promulgated by Minneapolitans, no doubt) that the first residents of St. Paul were forced to move there at gunpoint.

As the function of the fort changed after early settlement of the district around the Falls of St. Anthony, downtown St. Paul, and downtown Minneapolis, the reserve was redefined to open up land for urban development.

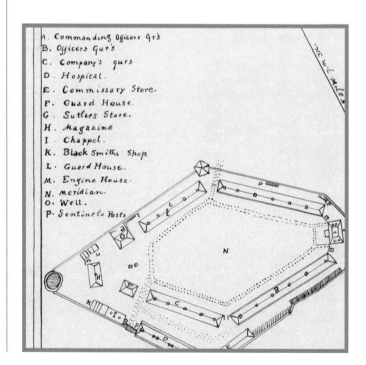

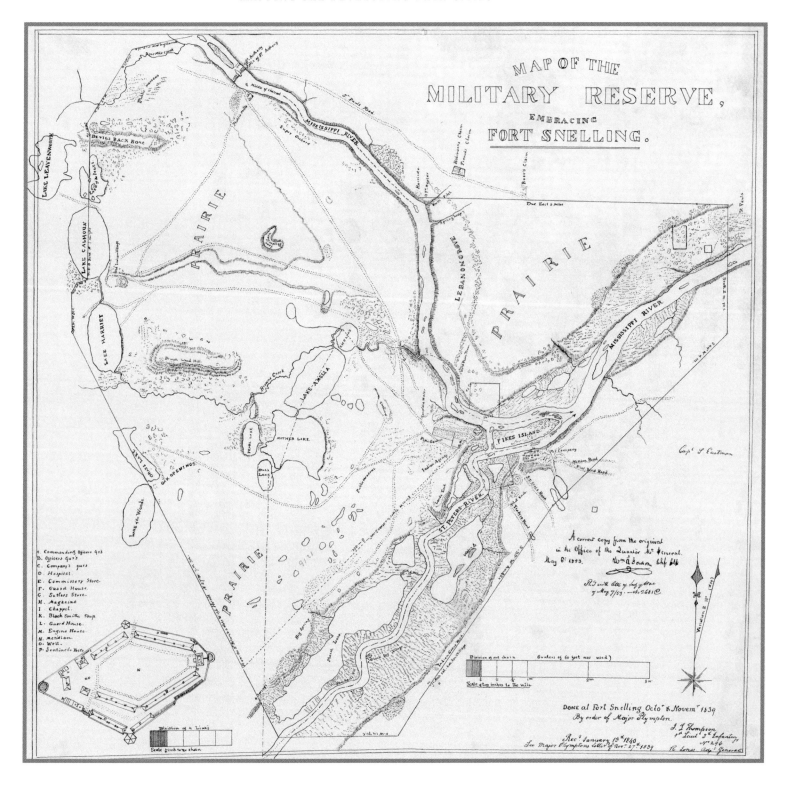

Geo. H. Ellsbury. *St. Paul, Minnesota. Lithographed by Hoffman.*

CHICAGO: CHAS. SCHOBER & CO., 1874

16.5 X 29 INCHES

MINNESOTA HISTORICAL SOCIETY

This complex and skillfully rendered view of St. Paul is the work of George H. Ellsbury, a New Yorker who moved to St. Charles, Minnesota, in 1857 at age seventeen. Although the portion of the map occupied by the cityscape across the Mississippi River is very small, it is superbly rendered. Prominent buildings are clearly shown, even if only by the tip of a steeple peeking over a warehouse.

Ellsbury took many artistic liberties when producing his view. Favored buildings are shifted to be in the line of sight, and their best facades are shown. Although he straightens out the Mississippi slightly, he beautifully illustrates St. Paul's superb location. The Mississippi, which flows from left to right across the page, was eroding the base of the cliffs where several tributary valleys intersected with the main gorge. This meant the steamboat captains could tie up next to the center of commerce and relatively easily send their cargoes up the hill. On the other (near) side of the river, the river was depositing silt and creating a large marshy expanse between the main channel and the high ground on which the artist stood. This situation made the development of a port on that side impractical.

Although harbors on both sides are shown full of activity, the far side hosts most of it. Eventually the upper levee (at left) would lose favor, and the railroad shown in the view would cut it off from the water completely. The lower levee would be taken over by the railroad depot and associated land uses. But on this sunny day, both forms of transportation are cooperating to make a great city.

Activity on the Mississippi River on this day in 1874 included three major shipping lines, as well as several aspects of the railroad operations that would in the next decade drive them all out of business. The view also clearly showed the basic structure of a large river town. The main street ran parallel to the river, and at the point where it was intersected by the bridge road, the center of commerce, government, and culture emerged at Bridge Square.

Although the town was only three decades old, substantial commercial structures and churches that identify the city's ethnic diversity had sprouted up. Growth in population and commerce caused a series of government buildings to be erected. By 1874, the Federal Custom House, the State Capitol, and the City Hall and County Court House had all been rebuilt

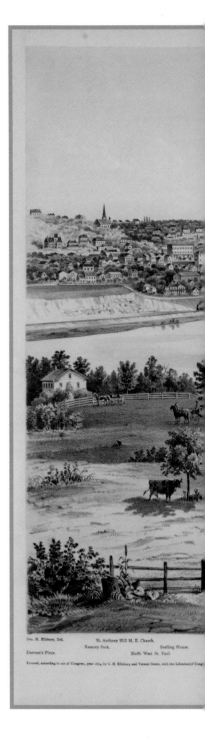

as splendid large structures and are appropriately featured. Like all successful cities of this era, St. Paul boasted a few modern hotels to provide guests with private rooms, comfortable beds, and a status address: the Metropolitan in its downtown location and the Park Place Hotel proudly flying the American flag and dominating the outlying area. The crown jewel of any proper city's skyline was the opera house, but St. Paul's is barely visible in the crowded blocks near Bridge Square. Unfortunately for the elegant citizens taking the view on the fine September Sunday, within six years Minneapolis would become the largest city in the state and attract most of the Twin Cities' new commercial developments.

Plan of St. Anthony Park, a Suburban Addition to St. Paul and Minneapolis.

CHICAGO: CLEVELAND AND FRENCH, 1873

10 X 15 INCHES

RAMSEY COUNTY HISTORICAL SOCIETY

In the northwest corner of St. Paul is a secluded neighborhood, separated from the rest of the city by the St. Paul Campus of the University of Minnesota, the State Fair Grounds, and an industrial area with easy access to railroad tracks. Its history is unique, but it has features that are representative of the development of all neighborhoods. Some parts of the city were continuous extensions of the original set of grids platted in the city's first years, but St. Anthony Park and several other areas began as suburbs totally outside the two core cities.

St. Anthony Park's original developers expected that their development would become a home for the wealthy residents of the town of St. Anthony Falls, who wished to escape its industry. The company of developers hired Horace W. S. Cleveland, one of the era's premier landscape architects, to design their community. Relying on his principles that a place's natural beauty and setting should structure the design, he produced a handsome residential plan for the district based on 5- to 25-acre lots, drawn not on a grid but sited in ways to maximize the scenic potential of each house site. Natural vegetation was to be supplemented by careful plantings, and home sites were to be connected by curved streets that would promote leisurely travel and help maintain a tranquil lifestyle. The intent was a picturesque community connected to the cores of both large cities by rail.

Unfortunately for the developers and for Cleveland, the terrible financial panic of 1873 ruined both their plans and Cleveland's practice. The land passed on to others, who formed the St. Anthony Park Company, led by a former Congregational minister named Charles Pratt. He was one of many easterners who came to Minnesota to improve their health and make a fortune in the process. Pratt and his associates also had a dream for a perfect suburban community, but their intended clients were middle-class families headed by men who managed or worked in the manufacturing firms that were expected to locate along the railroad tracks adjacent to the new residential areas. In 1885, the group filed another plan for subdividing the area. In the new plan, small lots dominate, although the streets are laid out just as Cleveland would have wanted, to respect the area's hills and marshes. Only fragments of Cleveland's original design are visible in St. Anthony Park today, but Cleveland later played a critical role in the creation of the parks and scenic drives linking Minneapolis and St. Paul.

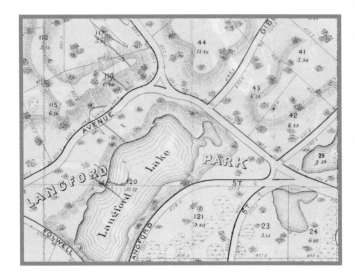

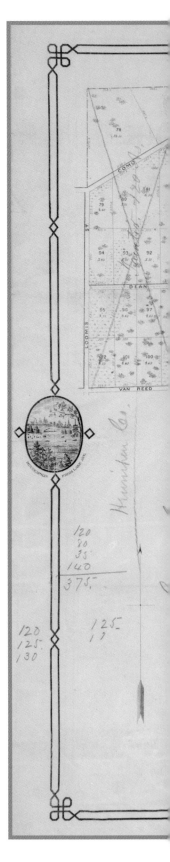

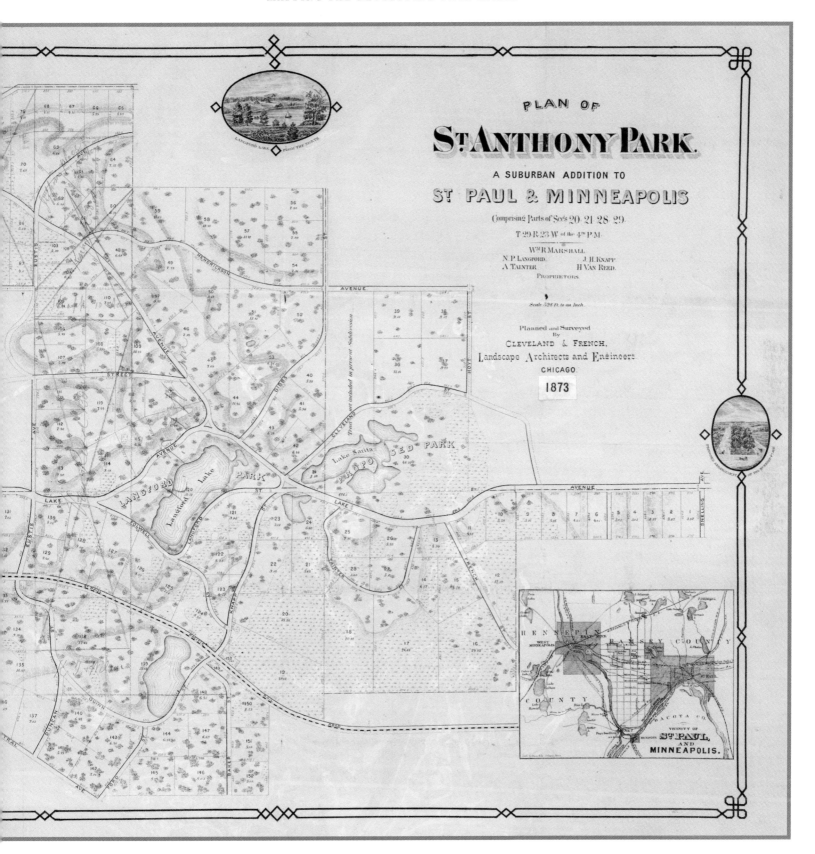

Report of the Leading Business Houses of Minneapolis, Minnesota.

MINNEAPOLIS: AUGUSTUS HAGEBOECK, 1886

10.8 × 16.4 INCHES

MINNESOTA HISTORICAL SOCIETY

Augustus Hageboeck's 1886 engraving of the Falls of St. Anthony district of Minneapolis shows it as it approached its years of peak milling production. Large industrial and commercial buildings line the riverbank, and the city extends away from the falls for several miles. The milling companies also developed a series of diversion tunnels to channel river water under the mills and power their turbines. James J. Hill's Stone Arch railroad bridge, which crossed the river in a sweeping curve, provided an efficient and functional testament to the new connections between the falls and Minneapolis, for it was railroads that supplied raw materials to the mills and distant markets for their products.

The wild cataracts of the Falls of St. Anthony were harnessed by the city during the 1870s and 1880s. Mills along the west bank stretched about half a mile from Fifth to Eighth Avenue, because the channel under First Street brought water to the mills to operate their turbines and because it was no longer necessary to be located on the falls themselves to use the waterpower. One set of mills faced the rails near the Milwaukee Depot; the others made use of tracks built on a viaduct over the river's edge.

By the mid-1880s, flour mills had pushed out the lumber mills that had previously dominated the site upstream. The imposing Pillsbury A Mill, which loomed over the east bank, replaced a sash and door company in 1881. Sawmills could still be found on the islands and over the east channel of the river, but they were soon to join the brickyards and other lumber-related businesses upstream toward the Camden neighborhood. The mills were making Minneapolis rich, and to let the world (and St. Paulites) know just how important the city was, real estate developers constructed a huge industrial exhibition hall upstream from the Pillsbury mill that proclaimed Minneapolis the center for all things industrial, urban, and modern.

The rapid ripening of the grain milling industry at the falls resulted from the growth of the railroad, the accessibility of nearby land, innovations in milling technologies, and creative marketing for the finished product in the United States and abroad. In order to produce an adequate supply of wheat, the millers worked with the railroad corporations that owned vast land grants west and north of the Twin Cities and in the Red River Valley.

By 1892, industrial Minneapolis was no longer dependent on water from the falls because it powered machines with steam generated by coal instead of falling water turning turbines. Industrial sites along the railroad tracks away from the river valley were developed for new industries and others relocating from the old district. A century later, the Falls of St. Anthony district would be redeveloped as a residential and retail area.

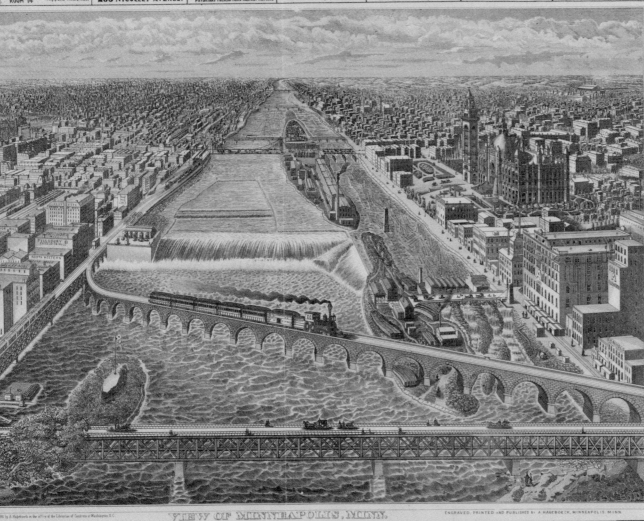

Lowry Hill,
from Insurance Maps of
Minneapolis, vol. 3.

NEW YORK: SANBORN MAP CO., 1912

27.5 X 39.5 INCHES

MINNESOTA HISTORICAL SOCIETY

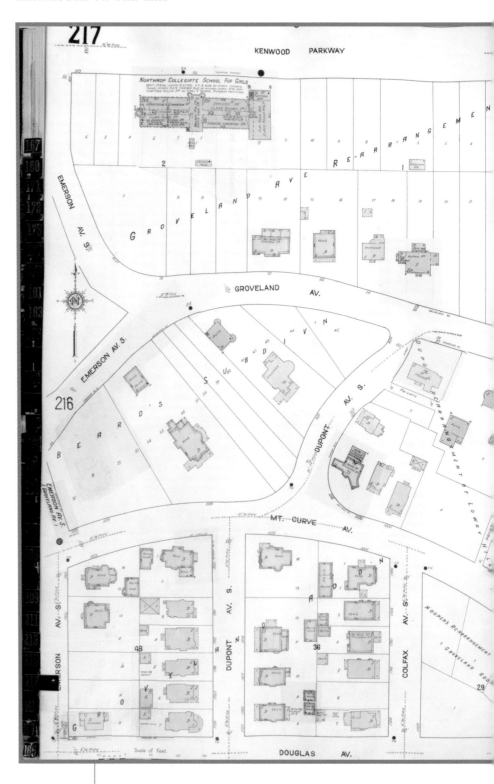

Insurance atlases of the nineteenth and early twentieth centuries, which first appeared around 1850, offer an unparalleled opportunity to examine details of the built urban landscape. They were created to compile current, accurate, and detailed information for insurance companies because companies needed to estimate the risk of fire and establish appropriate insurance rates. Underwriters working for national corporations could not personally inspect the properties to be insured, and therefore publishers like the Sanborn Map Company stepped in to supply the demand for maps that provided this essential information about the degree of hazard associated with a particular property.

Most of these maps are on large sheets at the scale of 50 feet per inch. Commercial and public buildings are frequently named. The fire-prevention infrastructure—fire departments, water and gas mains, and distances from fire hydrants to buildings—are highlighted. The outline, or footprint, of each large building is indicated, and the buildings are color coded to show the construction material (pink for brick, yellow for wood, and brown for adobe). Numbers inside the lower right corner of each building indicate the number of stories in the building; the numbers outside on the street front give the street addresses. The large maps were lithographed and then hand colored.

The five-volume Sanborn atlas of Minneapolis, which included selected suburbs, holds a treasure trove of information about neighborhoods and commercial districts in the

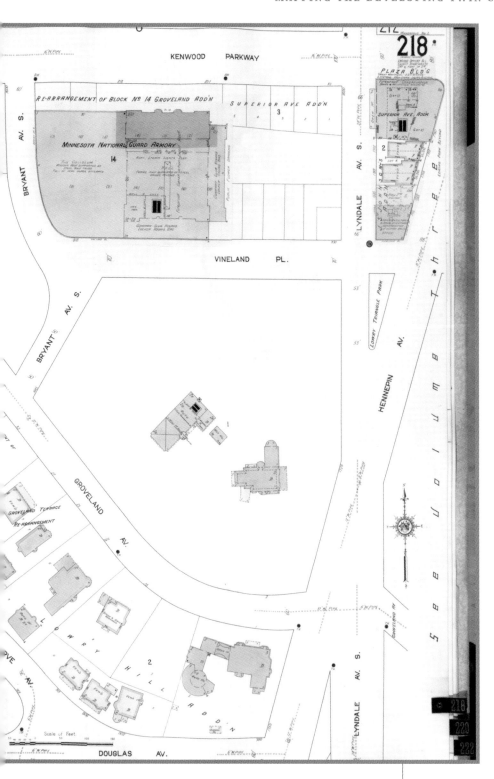

making endeavors. At the center of the development was the 4.35-acre hilltop estate built in 1874 by Thomas Lowry, head of the streetcar system, with his father-in-law, a major real estate developer. The house was three stories tall and built in the Empire style, with a mansard roof, bay windows, a grand porch, and plenty of chimneys. In 1915, the lumberman Thomas B. Walker purchased the house, and in 1927 he opened an art gallery there for his collection. (Today's reconstructed Walker Art Center sits on the site.)

Downslope from the mansion were the tightly packed residential and commercial spaces in the triangular block at the confluence of Hennepin and Lyndale avenues and the National Guard Armory and associated parade grounds. Northrop School—now Blake School—is shown as a large brick building occupying that site.

Groveland Avenue, now Groveland Terrace, developed as a stylish residential district in the 1890s. The largest houses were the enormous mansion at 1 Groveland Terrace, built by George H. Partridge, and the Romanesque home of the department store developer William L. Donaldson at 21 Groveland. At 15 Groveland, the Nott house was built by the architects Long and Kees, also in the Romanesque style. The prosperous Frank Long built his own house at 25 Groveland. The home of the Reverend Henry Beard, one of the first residents of the area, was at 1106 Mount Curve.

By the time this map was made, cars had replaced the horses and carriages from the beginnings of the neighborhood. The only stable still standing was at 1710 Dupont. This neighborhood would undergo considerable in-fill development and dramatic increases in density in coming decades. Here it appears as its developers intended it.

city. The Groveland Addition, of which Lowry Hill is a part, was platted by a consortium of men engaged in real estate development, railroad construction, milling, and other money-

Downtown Minneapolis,
from *Insurance Maps of Minneapolis,* vol. 3.

NEW YORK: SANBORN MAP CO., 1912

29.5 X 17 INCHES

MINNESOTA HISTORICAL SOCIETY

Insurance agents using the Sanborn atlas regularly received map updates designed to be pasted directly onto the older map pages. Alexander Dalrymple and other makers of marine charts had used this method of correcting working maps as early as the 1700s. There were few updates in residential districts once the first sets of houses were constructed, but commercial and industrial areas constantly changed. Maps of central cities used different scales and went into great detail in so-called congested areas. The maps were also layered with revisions as buildings changed functions or were renovated or replaced. These two adjoining Sanborn maps of Minneapolis capture the dynamic nature of the central business district and provide a glimpse of the process of abandonment as well.

Minneapolis's early downtown was Bridge Square, near where Hennepin Avenue crossed the river. As the city expanded and diversified, the business community moved into newer structures to the southwest. These buildings were more conveniently located for wealthy shoppers (who may have lived on Lowry Hill) and managers of the city's commercial activities. By the time these Sanborn maps of Minneapolis were made, the city center had already moved beyond this location. The block of buildings between Nicollet, Hennepin, Fourth, and Fifth illustrates not only the com-

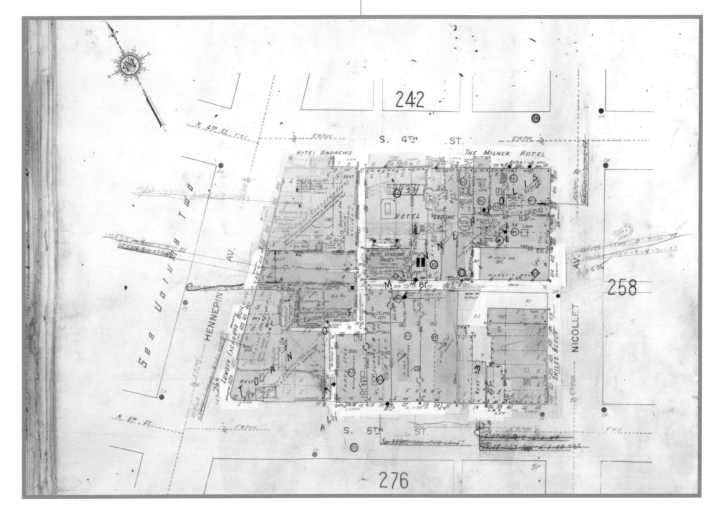

plexity of land use in the pre-auto central business district but also the large number of people who lived or stayed in the area. The Lumber Exchange building recalls the time when this area was the center of decision making in the local and regional economy, but by 1912 the area had been turned over into hotels, small storefronts, and an impressive number of small restaurants. The second floors of several buildings are vacant. The insurance company paid close attention to those sorts of businesses that were prone to fire and so identified restaurants, gas stations, garages, and functions that might cause trouble. (Other types of businesses were not identified.)

On another downtown Minneapolis sheet (258), Powers Mercantile department store, one of several mainline fashion establishments, sits across Nicollet. Department stores, like office buildings, were made possible by the electric streetcar and the elevator. Streetcars enabled all sorts of middle-class residents to shop and work downtown. Elevators enabled workers and shoppers to easily move through the highrise buildings that were increasingly dominating American cities. In addition to shopping and corporate offices, Minneapolis's downtown offered financial services, including Farmers and Mechanics Bank, one of the financial anchors of the community. The map also shows several garages that have been built in the interior of the blocks. A paste-over update from the late 1940s shows that the area was being converted into an automobile landscape. Much of the downtown core shown in these maps would experience urban renewal in the 1960s and 1970s when the city leaders set out to create a new, vibrant, and modern downtown devoid of the reminders of times that had passed.

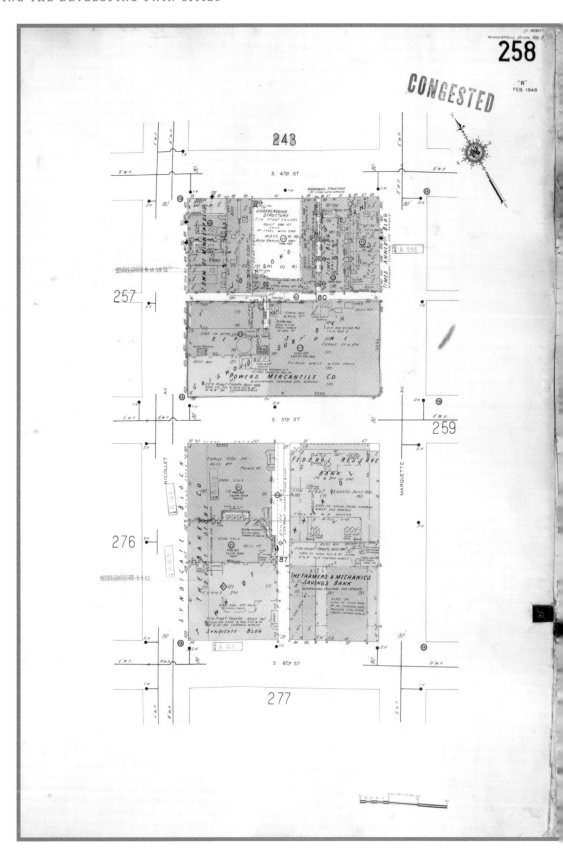

Section 32, Township 29, Range 22, plate 14, from Atlas of the City of St. Paul.

CHICAGO: RUBEN H. DONNELLEY, 1892

29.5 X 34 INCHES

AUTHOR'S COLLECTION

Maps sometimes depict the heights and depths of a city, both geographically and economically. This view of a portion of St. Paul just east of downtown shows the low-lying squatter settlement of Svenska Dalen, or Swede Hollow, along the Phalen Creek bottom. High on the bluff overlooking the Hollow sits the grand mansion of Theodore Hamm, successful owner of St. Paul's premier brewery. Overall, railroad tracks, railroad yards, freight houses, and engine houses dominate this plat.

During the late nineteenth century, Swede Hollow was a "stepping stone" neighborhood, a place where impoverished newcomers could find cheap shelter and a refuge until they could afford better housing. Many Swedes eventually settled out into the city's East Side neighborhood.

Unlike the residential areas surrounding it, Swede Hollow had no regular grid pattern of streets and alleys, and the only street mapped in the Hollow was really a country lane that meandered along the side of the creek. Houses were truly vernacular—small, with many tiny additions, and made from lumber salvaged from construction sites. The dwellings sat close to each other on whatever space was available. Residents took water from a natural spring, used the creek as their sewer, and maintained gardens and barnyard animals. The absence of the property lines that dominate the rest of the city indicates that occupants did not own the land. It belonged to the power company, which viewed it as a potential waterpower site that could be flooded when necessary. In spring, the rising creek encouraged a massive neighborhood cleanup, and anything unwanted was thrown into the flood tide.

Beginning around 1881, immigrants from Ireland joined the Swedes in the Hollow. Relations were not always amicable. Years later, one resident, Nels M. Hokanson, remembered that "combative Irish boys threw stones at the drum during the Salvation Army services" and "picked fights with Swedes and harassed me at every opportunity."

A series of immigrant groups followed the Swedes and the Irish through Svenska Dalen after 1900. Italian became the dominant language on the lower East Side, and the Hollow was known for the smell of fermenting grapes each fall. Mexicans followed after World War II. In 1956, the spring that had served as the community's water supply for over a century was declared unsafe. The last fourteen families occupying Swede Hollow were evicted, and all the houses were burned by the city fire department. Today, a sign at its entrance welcomes visitors to a historic forest.

Many former residents of the Hollow look back on their childhoods there with nostalgia, although public health workers who visited the area remembered the numerous cases of whooping cough, pneumonia, and undernourishment. Survivors were tough.

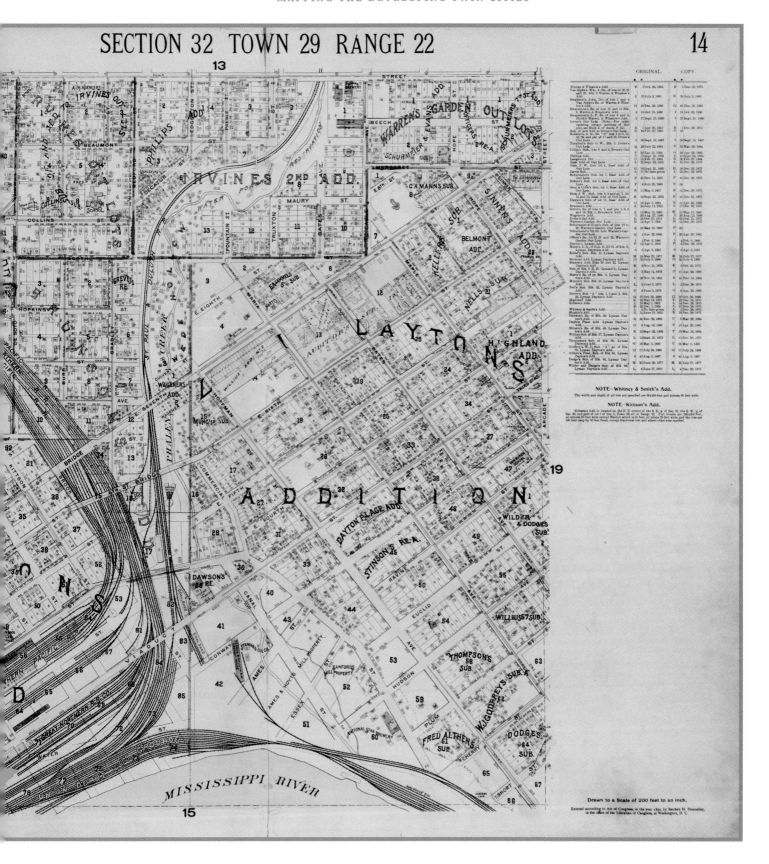

St. Paul, Township 28 North, Range 22 West, Plate 52, from Hopkins Plat Book of St. Paul and Suburbs.

PHILADELPHIA: G. M. HOPKINS CO., 1926

29.5 X 17 INCHES

DAKOTA COUNTY HISTORICAL SOCIETY

In the 1870s and 1880s, a small stockyard and two small slaughtering facilities sat along the Minnesota Transfer tracks in the Midway District of St. Paul. This operation functioned primarily to provide water and rest for livestock on their way to Chicago and to supply the local demand for meat. When the investors in these corporations would not expand their facilities to meet the growing demands of the ranchers on the northern plains, Alphas P. Stickney, president of the Minnesota Northwestern (later the Chicago–Great Western) Railroad, saw his opportunity. Buying options on land along the Mississippi River three miles downstream—and downwind—from St. Paul, he and his associates planned to build a new union stockyard based on the model of Chicago. These two plat maps present slightly different parts of the site Stickney selected and of the housing lots to the west of the yards.

All the developers needed was a secure supply of livestock. In spring 1886, Stickney convinced the members of the Montana Stockgrowers Association at their meeting to send their animals to the South St. Paul yards. Working in partnership with J. J. Hill, Stickney persuaded the St. Paul Chamber of Commerce and other railroad investors and banks to back the plan. In June 1886, the Union Stockyards Company was incorporated, financed in part by bonds sold in England. The town of South St. Paul was incorporated a year later to house the workers and provide services for the huge development on the river flats. The railroad-oriented complex grew with the expansion of the livestock industry in the Upper Midwest and the northern plains.

This South St. Paul site had several advantages. It was served by two railroads and was adjacent to the river. Ice was a critical resource in the packing industry, and, as seen on the

map, huge icehouses were built to keep the ice through the summer. In addition, several artesian wells on the site provided a nearly unlimited supply of high-quality water for the animals and the processing of meat. Because the site was very swampy, the nearby river bluffs were excavated and the river bottom dredged for fill to cover the 170-acre site with 5 feet of sand.

The actual stockyards consisted of a grid pattern of pens and alleys, constructed on a brick pavement. The pens in the cattle division were enclosed with fences 8 feet high, with 2-by-6 planks on both sides of the posts. The fences were topped with two rows of planks laid side by side so workers could easily walk along the tops of the fences to move from pen to pen. The stout fences could contain even the strongest and wildest bulls from the western ranges. Fences in the hog division were built on the same design but were 4 feet tall. A series of gates enabled workers to establish routes through acres and acres of pens.

After a decade of considerable uncertainty, the operation's future solidified when Swift & Company began to expand. After opening new plants in Kansas City, Omaha, and St. Louis, Swift signed a 99-year lease and took over the operations of the failed Minnesota Packing and Provision Company. Both the hog and the cattle-kill operations were expanded, and a new smokehouse was built. All are depicted on the maps. The office building facing the main road leading into the complex is dwarfed by the huge processing and storage facilities.

In 1900, eleven major corporations employed 558 men here, and by 1905 the number of workers had increased to about 700. But the facility's greatest growth was still in the future. Demand for food during World War I prompted the expansion of Armour and Company, the second great meat-packing corporation. In November 1919, after spending over

$14,000,000 on construction and site preparation, Armour and Company opened its ultramodern, fireproof, and highly sanitary slaughtering and packing facility. This brick complex overshadowed the eclectic collection of older buildings operated by Swift and the other packers. When the Armour plant opened, it added 2,000 to 3,000 jobs. At its peak in 1941, the entire complex provided jobs for 8,000 people and processed over 4,250,000 animals.

The packers bragged about using every part of the hog but the squeal. They found a use for all parts of the animals, but in order to market this complex set of goods, the companies needed to make their own packaging. Thus, the maps show box factories in the complex, as well as smoking and curing buildings. The best meat was sold fresh, and beef-luggers carried halves of dressed cattle, hog, and sheep carcasses to lines of refrigerator cars, which were cooled with ice. Once loaded, the meat train would "highball it" to the East Coast to get the meat to market before the carcasses spoiled.

Leftovers such as hides went first to the hide cellar, an abominable place, where they were soaked and cured before going to the tannery to be made into leather. Bones were used to make ink and, in later years, bone meal. The meat that could not be used in a recognizable standard cut became sausage. The scraps and damaged or contaminated cuts became pet food. In addition to the meat market, there was a market for horses and milking cows here as well. These facilities sat outside the main stockyard complexes adjacent to Swift's.

During the 1960s, meatpackers decentralized their operations throughout the Midwest, and the older plants were closed. The Swift plant in South St. Paul closed in 1969, and Armour closed in 1979. Due to advances in roads and interstates, meat processing moved closer to the suppliers. The large complexes of Swift, Rifkin, and Armour have been demolished, and the new businesses that replaced them have no connection to the historic uses of the site.

The Twin Cities: Their Famous Lakes, River, Parks and Resorts, from *Annual Report of the Twin City Rapid Transit Company.*

MINNEAPOLIS: TWIN CITY RAPID TRANSIT COMPANY, 1909 (FACSIMILE; MINNEAPOLIS: UNIVERSITY OF MINNESOTA, 1989)

10 X 33 INCHES

MINNESOTA HISTORICAL SOCIETY

This banner-like, poster-sized map has a most disingenuous title. Although it shows the lakes, river, parks, and resorts, it is really intended to advertise the local streetcar company. The bright red lines marking the track and boat routes grab the viewer's attention, to the exclusion of the other features of the map. It is neither a map nor a true bird's-eye view, but it shares some elements of each.

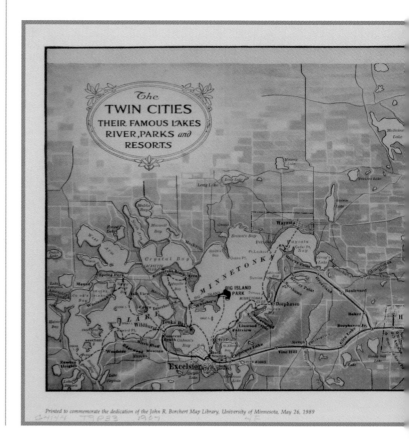

MAPPING THE DEVELOPING TWIN CITIES

The length of the map has been distorted to exaggerate the east-west distance, which emphasizes the wide expanse covered by the streetcar system. The muted background sketchily suggests land uses and roads and contains a few landmarks, recreation sites, and major inner-city places of employment, such as the flour mills at the Falls of St. Anthony, the lumber mills on the river bank in northeast Minneapolis, and government buildings. Because the street railway was a metropolitan system that extended across Minneapolis and St. Paul, the map also highlights suburban places of employment. The streetcar management was especially proud of the maritime portion of the system. The major steam lines on Lake Minnetonka are shown, along with the much smaller excursion routes on Lake Harriet and White Bear Lake.

The development of a streetcar system in the Twin Cities fostered all manner of changes along the network. At first the system had two relatively independent downtown centers on which the lines in each city focused. The lack of lines connecting the two cities, in part a function of limited river crossings, helped maintain their separate identities and the rivalry. Each downtown was the destination of thousands of passengers who worked, shopped, and entertained themselves in the offices, department stores, and new movie palaces. Away from the downtowns, the tracks provided a linear structure for the development of new neighborhoods that tended to be rather homogeneous in style and building value. At the

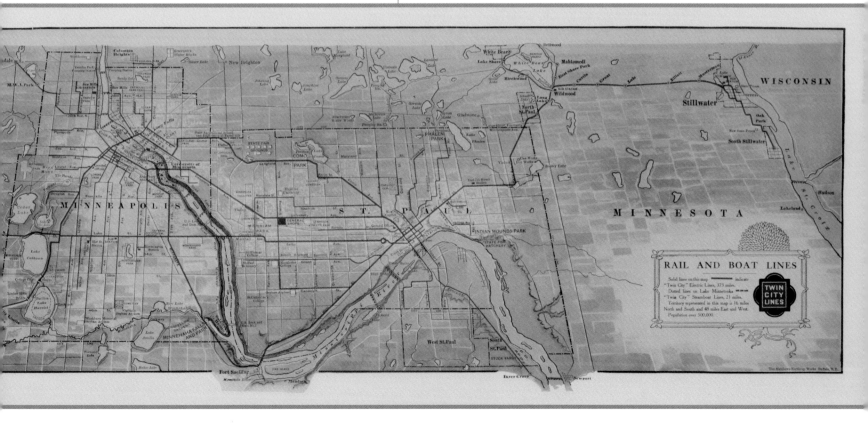

intersections of streetcar lines, small commercial districts grew up, and, during the 1920s, large brick apartment blocks and six-plexes were built on several lines. Service was quite good. During the three rush hours, cars ran at 5-minute intervals on the more commercialized streets. The cars carried mail, newspapers, and parcels. In an era when many men chewed snoose or tobacco, it was necessary to post signs admonishing passengers not to spit on the floor. Smoking was allowed on the back platform, however.

For a time, the company also owned three large ferryboats to take revelers from the stations to Big Island Amusement Park. The ferries carried 1,000 people each and are reputed to have regularly disembarked 15,000 passengers on Big Island on Sundays and holidays. By 1911, the novelty had worn off, and Big Island Amusement Park was sold. The streetcar system fell into disrepair and lost its attractiveness during the years after World War II. Everyone of importance, or who thought they were important, wanted a car and did not want to ride mass transit. To make matters worse, the rather conservative but profitable company was taken over by Charles Green of New York and Fred Ossanna from the Twin Cities. These two bled the company and, with the help of special financing by General Motors, replaced the streetcars with 525 buses in 1953.

Minneapolis West.

U.S. GEOLOGICAL SURVEY, 1901
(REPRINTED 1915)

20 X 16.25 INCHES

AUTHOR'S COLLECTION

This U.S. Geological Survey map provides a fascinating glimpse of the great natural forces that shaped the surface of Minneapolis's landscape before its urbanization. The Twin Cities are nearly surrounded by a belt of low hills, or moraines, and by small lakes to the west, south, and east. Only a few hills, however, sit within the city limits of Minneapolis.

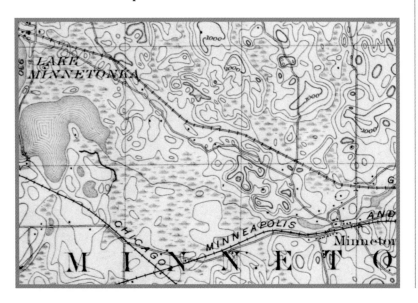

Lowry Hill, the Devil's Backbone, is probably the best known, but two other important elevations stand on the east and west sides of Lake Calhoun. West of the city, a vast area of swamps, moraines, and a few small lakes stretches toward the very large glacially formed Lake Minnetonka. Minnehaha Creek flows out of Minnetonka and meanders across this very young landscape, eventually reaching the Mississippi. Nine Mile Creek and Purgatory Creek also drain the large marshes occupying the spaces among the moraines north of the Minnesota River. This rolling landscape supported a forest of hardwood trees, mixed with some conifers, before the coming of farmers.

By 1900, the pace of industrial development in Minneapolis had outstripped the capacity of the old river-oriented industrial districts. The lobe of development that extended toward the southwest from the core was particularly important in the growth of the city's economy and culture. The backbone of this wedge of growth was the network of railroad tracks and roads that fanned out from central Minneapolis heading for the rich agricultural land on the prairies beyond the belt of low hills in western Hennepin County. The railroad tracks were squeezed into a corridor, which was even-

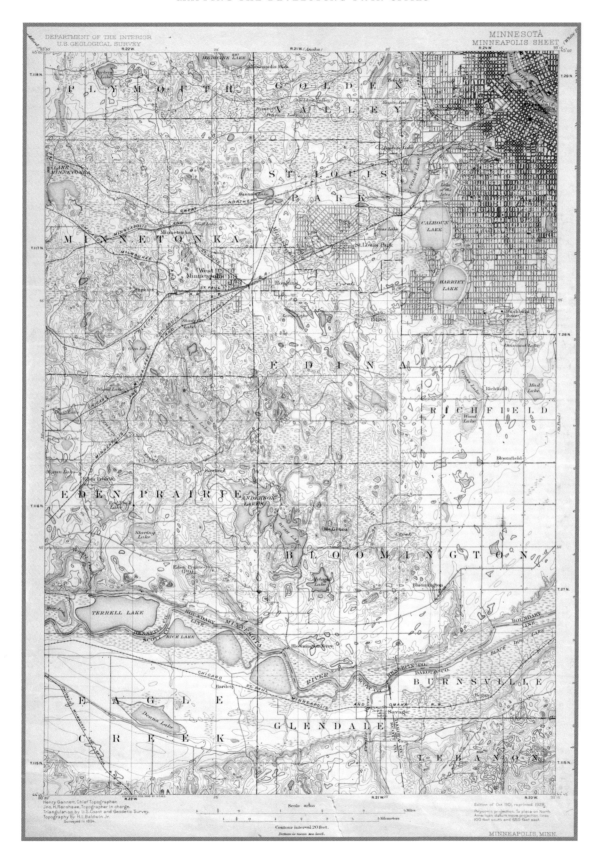

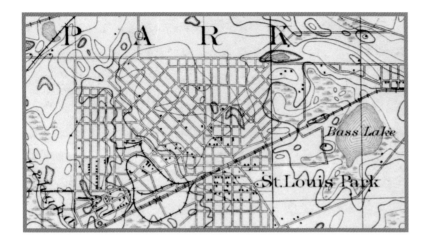

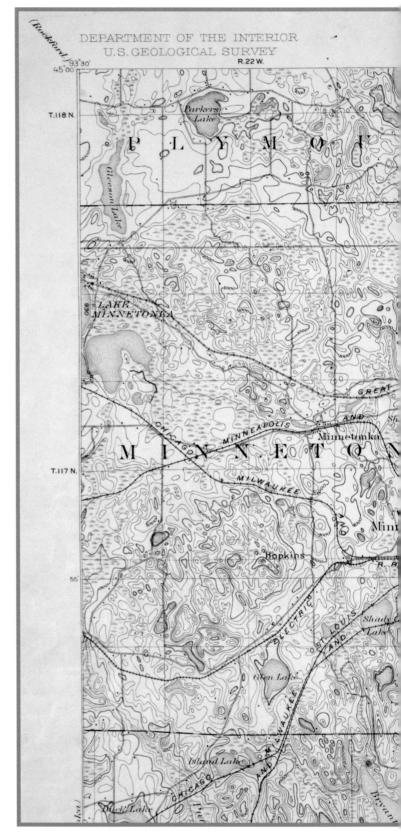

tually compressed into a tight band of tracks, with heavy traffic passing north of Lowry Hill on the way to the river crossing at Nicollet Island and on to the east. The big loop of tracks in St. Louis Park indicates a major factory development. In the coming years, Hopkins's location at the intersection of several rail lines would attract the Minneapolis Moline farm implement factory.

By this map's date, 1901, the beginning of the city's spread is marked by the platted streets, but without the small black squares that symbolize detached houses. The homes of the city's elite on Lowry Hill have captured the stunning view of downtown. Farther along the brow of the hill, in the Kenwood development, a recognizable cluster of homes has been built. The marshy western shore of Lake of the Isles is undeveloped, as is the western side of Lake Calhoun. Most of Golden Valley Township is a vast marsh and unoccupied. The streets east of Lyndale and north of Lake Street are lined with houses, but close to the lakes are scattered clusters of houses. Finally the grid ends, and occasional houses line some of the township roads in loose ranks.

Because the land outside the city in the Minnehaha Creek watershed was extremely marshy, the old roads meandered through the area following the high ground. The Edina Post Office is shown, but this crossroads settlement is dwarfed by its much larger neighbors to the north—St. Louis Park, West Minneapolis, and Hopkins. In years to come, the marshes were drained and filled, and many of the hills were excavated to make way for more suburbs.

Detail, Minneapolis West, 1901 (page 151)

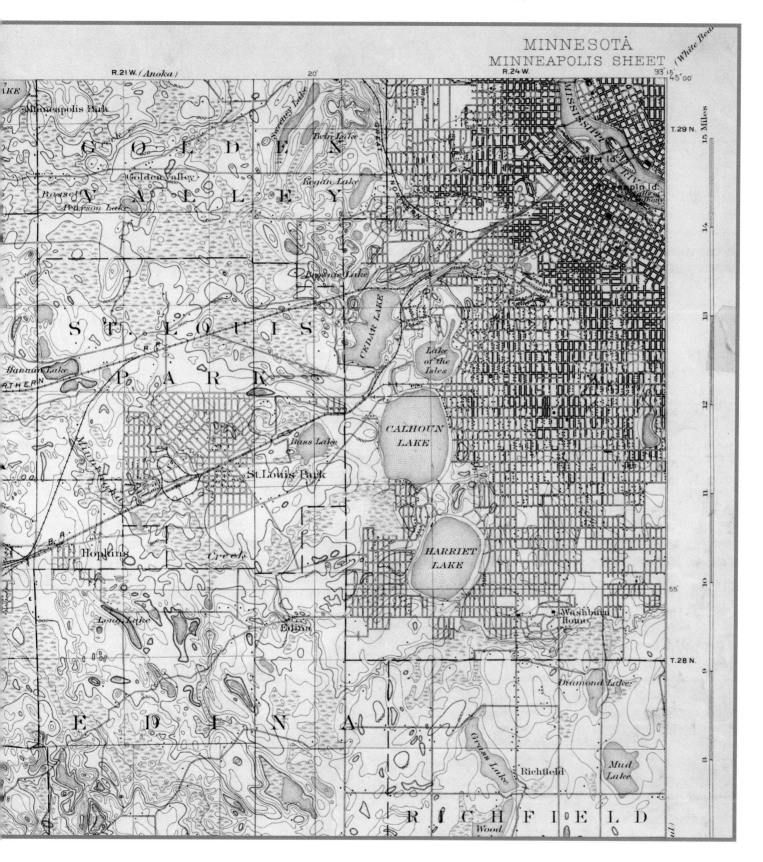

Map of Minneapolis, Minnesota, Showing Park System as Recommended by Prof. H. W. S. Cleveland, from H. W. S. Cleveland, Suggestions for a System of Parks and Parkways for the City of Minneapolis, Read at Meeting of the Minneapolis Park Commissioners, June 2, 1883.

MINNEAPOLIS: JOHNSON, SMITH AND HARRISON, 1883

8.7 × 10 INCHES

MINNEAPOLIS PUBLIC LIBRARY

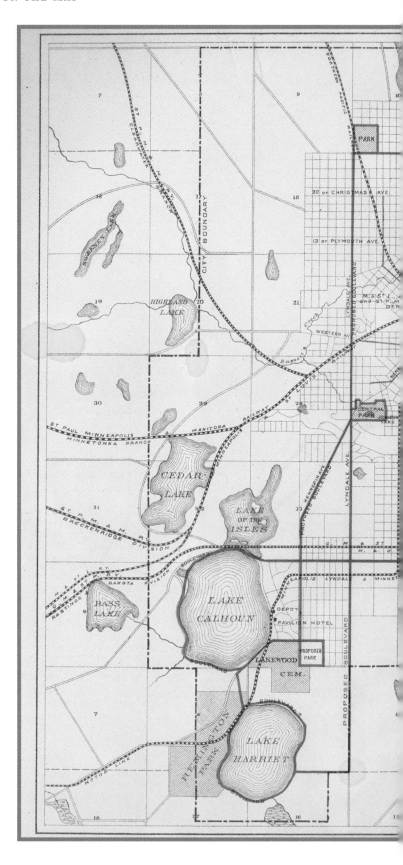

Although leaders in both cities discussed the possibility of establishing parks shortly after statehood, only a few squares of land were set aside in the central cities and wealthy neighborhoods, and no comprehensive plans were prepared. Then, in the winter of 1872, Horace Cleveland arrived to give a lecture titled "Landscape Gardening as Applied to Works of the West." His presentation caused a great stir, and in 1873 he was hired by both cities to design a series of parks and parkways. It took Minneapolis leaders a few years to get a park system established and to create, in 1882, the Minneapolis Park District, which had an independent board and the power to levy taxes. In 1883, board leader Thomas Lowry hired Cleveland to prepare a formal plan.

Cleveland's system, based on boulevards rather than on small parks, envisioned vistas being kept as natural as possible. He proposed the creation of boulevards around the lakes and an Interlaken Park between Lakes Calhoun and Harriet that would knit together a series of larger parks. The "priceless jewel" of the Mississippi River gorge especially impressed him.

After the park board accepted the plan and began to acquire land, Theodore Wirth, the Park District's second superintendent, augmented the plan. Wirth oversaw the creation of the parkways and the dredging and linking of the lakes. Cleveland had promoted the preservation of the entire river bluffs with a trans-Mississippi park along the gorge and a large pair of parks opposite each other at the confluence of the Mississippi and Minnehaha Creek on the St. Paul and Minneapolis

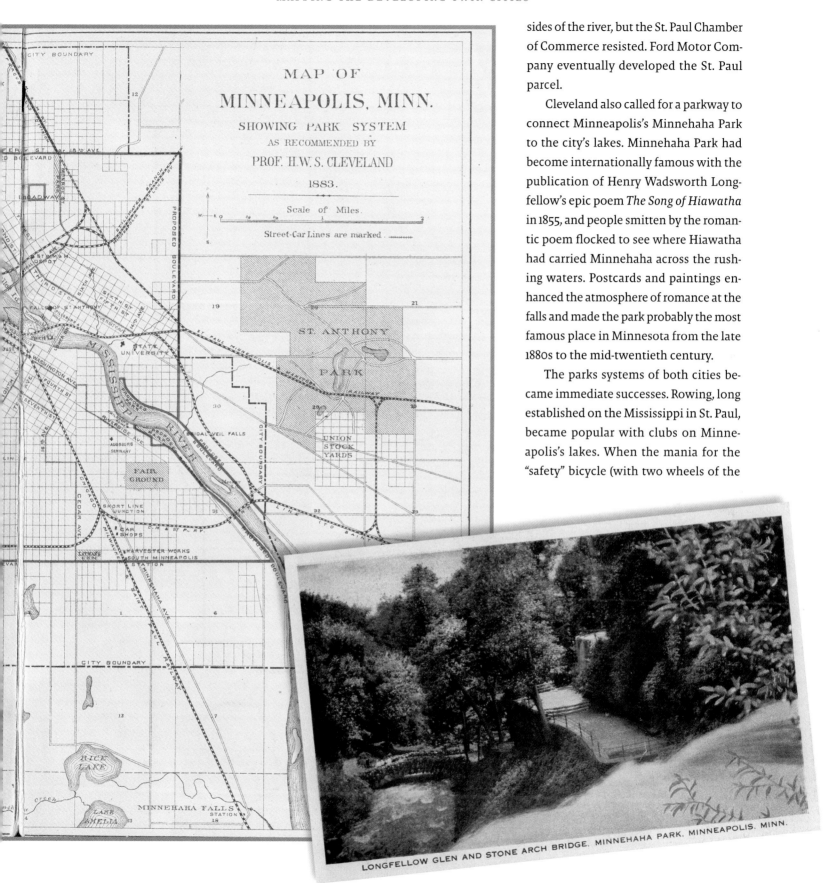

Longfellow Glen and Stone Arch Bridge, Minnehaha Park, Minneapolis, Minn.

sides of the river, but the St. Paul Chamber of Commerce resisted. Ford Motor Company eventually developed the St. Paul parcel.

Cleveland also called for a parkway to connect Minneapolis's Minnehaha Park to the city's lakes. Minnehaha Park had become internationally famous with the publication of Henry Wadsworth Longfellow's epic poem *The Song of Hiawatha* in 1855, and people smitten by the romantic poem flocked to see where Hiawatha had carried Minnehaha across the rushing waters. Postcards and paintings enhanced the atmosphere of romance at the falls and made the park probably the most famous place in Minnesota from the late 1880s to the mid-twentieth century.

The parks systems of both cities became immediate successes. Rowing, long established on the Mississippi in St. Paul, became popular with clubs on Minneapolis's lakes. When the mania for the "safety" bicycle (with two wheels of the

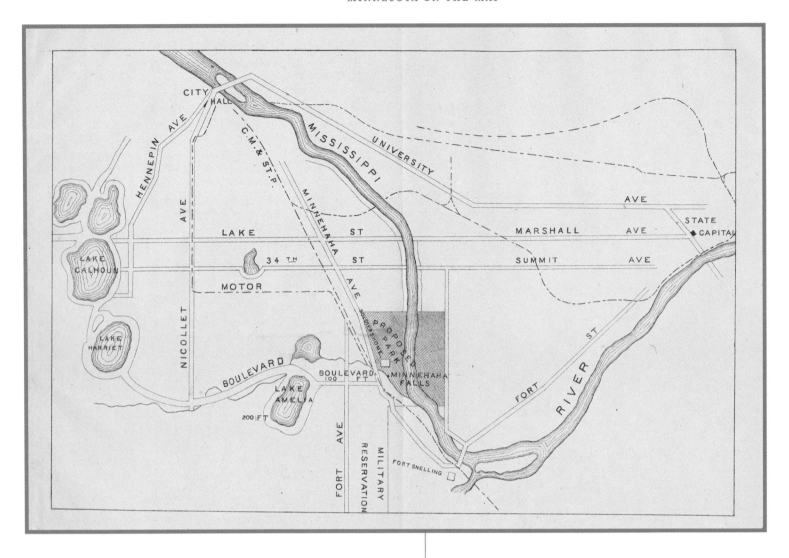

Map of Proposed Minnehaha Park and Parkways,
from *The Aesthetic Development of the United Cities of St. Paul and Minneapolis.*

AN ADDRESS GIVEN TO THE MINNEAPOLIS
SOCIETY OF FINE ARTS, APRIL 2, 1888

4 X 5 INCHES

MINNEAPOLIS PUBLIC LIBRARY

same size) arrived at the turn of the century from England, cyclists began speeding along the parkways from city to city. Parks also afforded sites for canoe clubs, a yacht club, and, perhaps most amazingly, horse racing on the frozen surface of Lake of the Isles. Streetcar traffic to the parks was especially heavy on weekends and on holidays, when special runs would efficiently carry people from their homes in the inner-city neighborhoods and the parks. The passing of time has brought many changes in uses of the beaches and parks, including races along parkways, but the fundamental beauty of the parks through the seasonal cycle continues to inspire and invigorate Twin Citians. It all began with maps!

Jules Guerin. *The Sixth Avenue Artery,*
from Edward H. Bennett and Andrew Crawford,
Plan of Minneapolis.

MINNEAPOLIS: CIVIC COMMISSION, 1917

8.25 x 4.6 INCHES

AUTHOR'S COLLECTION

The great White City of the Chicago World's Fair of 1893 captured the imaginations of visitors to the civic extravaganza along Lake Michigan or of those who saw photographs of the park that Daniel Burnham had created on Chicago's lakefront. Minneapolis's civic leaders were no exception.

Minneapolis's population growth since 1880 had been stunning. With a population just over 46,000 in 1880, the milling city grew by 1910 to over 300,000. Some (mistakenly) expected it to reach 400,000 by 1920. At any rate, by World War I, Minneapolis had won the population race with St. Paul and was clearly the larger sibling.

Minneapolis's leaders wanted to make their city into a great metropolis and believed that in order to achieve greatness the community needed both a vision and a plan. Because no city planning agency yet existed, a citizen committee known as the Civic Commission of Minneapolis was formed in 1910 for the purpose of developing and then promoting a plan for the city. The commission raised money and hired Edward H. Bennett and Andrew Wright Crawford from Chicago for the task. Prior to his death, the great architect and planner Daniel H. Burnham had been a partner with Bennett, and Burnham's influence is clearly seen in the 1917 *Plan of Minneapolis*. The Burnham and Bennett firm was the country's leading proponent of the City Beautiful movement. The firm had grand and optimistic visions for the future of urban life as the century of progress unfolded, and the men believed that a city did not have to be ugly to be efficient.

The Civic Commission of Minneapolis also fervently believed in a bright future for its city, but members were concerned about certain developments in the city. Traffic had become very congested in the downtown and on the major

roads. In some older parts of town, building deterioration was widespread. The planners' research had indicated that more growth in the neighborhoods was likely in response to the extension of streetcar lines. They also concluded that Min-

Jules Guerin. *The Sixth Avenue Approach to the Institute of Arts, through Washburn Park,* from Edward H. Bennett and Andrew Crawford, *Plan of Minneapolis.*

MINNEAPOLIS: CIVIC COMMISSION, 1917

6.5 X 10 INCHES

AUTHOR'S COLLECTION

neapolis's natural economic trade area was about as large as the country of France and that the Minnesota nerve center should have the grand look and feel of Paris. Their research supported three primary directions for renewal: construction of two great roads by widening and lengthening Sixth Avenue and Eighth Street and other extensions of the street system; creation of a municipal center at the junction of the two streets; and close linkage to the downtown of an art and education center at Fair Oaks Park and the Minneapolis Art Institute. In addition, a government center at City Hall and a transportation center at Gateway Park would be developed. Finally, the riverfront should be redeveloped. Clearly, this entire grand plan was strongly influenced by, if not derivative of, the Burnham-Bennett Plan of 1909 for Chicago.

Minneapolis's consultants presented the plan as a limited-edition book of 1,000 numbered copies printed on the highest-quality paper. Its 230 pages were lavishly illustrated with engravings showing elegant cityscapes from around the world and scores of maps and diagrams of the present and potential Minneapolis. The well-known artist Jules Guerin made these renderings, as well as the cityscape drawings, in his New York studio.

Critical to the plan was a grand boulevard. Sixth Avenue Parkway would be created by extending Sixth Avenue South (now Portland) southwestward from its turn to the north shore of Lake Harriet. This route would take it past the north facade of the Art Institute, ending at a dramatic ornamental gate, Watergate, about where the Rose Garden is now. Going the other direction, Sixth Avenue Parkway would run through downtown and cross the river along the line of Second Avenue to Central Avenue. From that point, one leg would follow Central north; the other would follow Hennepin Avenue to the city limits. The main section would bisect the angle and lead to a new garden suburb in what is now St. Anthony. In the downtown, it would lead to a two-block plaza. This parkway would ease the auto traffic congestion experienced by commuters driving in from the suburbs, facilitate movement in the downtown, increase the tax base by encouraging expensive new buildings along its length, and necessitate slum clearance of some blighted areas. It had the added advantage of providing a firebreak in the event of a conflagration.

Downtown's streets were to be widened, and land was to be purchased for the inevitable expansion of the government buildings. Height limitations would be imposed and skyscrapers quarantined in the section they "infested." A new union railroad station would be added to keep up with St. Paul. The planners recognized the riverfront as an "untouched municipal goldmine" and called for new bridges and parkways on both sides that would replace the blighted buildings. They further suggested that Nicollet Island above the falls be developed as an airport or a huge stadium. If the stadium were built, they argued, Minneapolis might be the site of a postwar Olympic Games.

Paris on the Mississippi? The planners' visions may seem unrealistic today, but their specific proposals are hauntingly familiar. The grand boulevard moving traffic from the suburbs to the downtown sounds something like I-35W. The notion that improvements would pay for themselves resounds as "tax increment financing." The idea of stadiums in the core city is current again, and, although today's developments along the Mississippi do not look like the ones envisioned in the plan, they have the same function.

The 1910 exercise in citizen planning for the city was not implemented for a host of reasons. The plan may or may not have been impractical, but it clearly failed to convince the leaders of real estate and property management companies to

THE SIXTH AVENUE APPROACH TO THE INSTITUTE OF ARTS, THROUGH WASHBURN PARK.

change their ways. Nor did it offer a great deal to the middle class, and so no broad base of citizen support developed for it. In the end, the beautifully produced plans were shelved and known mainly to bibliophiles. But Minneapolis has not yet disproved Burnham's famous maxim, quoted on the first page of the book containing the plan: "Make no little plans; they have no magic to stir men's blood and probably themselves will not be realized. Make big plans; aim high in hope and work, remembering that a noble, logical diagram once recorded will never die, but long after we are gone will be a living thing, asserting itself with ever-growing insistency. Remember that our sons and grandsons are going to do things that would stagger us. Let your watchword be order and your beacon beauty."

St. Paul Zoning Map, 1922,
from City Planning Board, Edward H. Bennett
and William E. Parsons, consultant city planners,
and George H. Herrold, city plan engineer,
Plan of St. Paul: The Capital City of Minnesota.

ST. PAUL: COMMISSIONER OF PUBLIC WORKS, 1922

17 X 23.5 INCHES

AUTHOR'S COLLECTION

In 1922, St. Paul's planning board submitted to its citizens a grand plan produced by the same firm that had created the City Beautiful plan for Minneapolis. Minneapolis's elegant planning volume was issued in a signed and numbered edition filled with color engravings, but St. Paul's was a modest black-and-white paperback filled with special-purpose maps. Both plans provided information about the nature of the city and called for new parks and rebuilt roads to handle rapidly increasing traffic. The St. Paul plan, however, introduced the new concept of land-use zoning and contains the city's first zoning map.

Since its mid-nineteenth-century founding, the state's capital city had been managed by real estate developers who platted neighborhoods and worked with the city's Department of Public Works to develop roads. In 1905, Cass Gilbert, the architect for the new state capitol building, had developed a set of sketches for aesthetic approaches to the capitol, and his plans were elaborated upon for later use in redeveloping the city as part of the City Beautiful movement. A city-planning ordinance passed in 1917 required all infrastructure development decisions to be approved by the planning commission. After World War I, a series of educational sessions and public meetings culminated in 1920 with the hiring of George H. Herrold to manage the planning process and the Chicago consultants of Edward H. Bennett and Andrew W. Crawford to develop the plan. Two years later, St. Paul possessed a highly modern urban plan, of which its zoning map is the most distinctive component.

English common law, the foundation of real estate law in the United States, protects property rights and generally enables property owners to do whatever they wish with their land. Zoning regulations, on the other hand, limit the activities that can occur on a piece of property so as to protect the health and welfare of the community. In essence, the new zoning laws enable local governments, using their police powers rights, to take development rights from property owners without compensation. American land-use zoning, which originated in New York City in 1916 as a reaction to the building of skyscrapers, became the model of law adopted by most states by 1920. Although the concept of zoning was challenged in the courts as a violation of the Fourteenth Amendment, the Supreme Court case of *Village of Euclid v. Ambler Realty Co.* (1926) firmly established the legality of zoning.

In Minnesota, the state legislature granted St. Paul the right to zone in 1921. The city approved, on July 7, 1922, a comprehensive zoning law and map to protect neighborhoods from encroachment

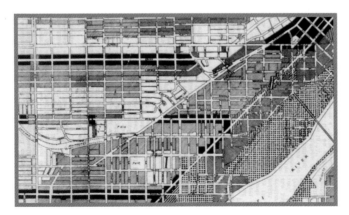

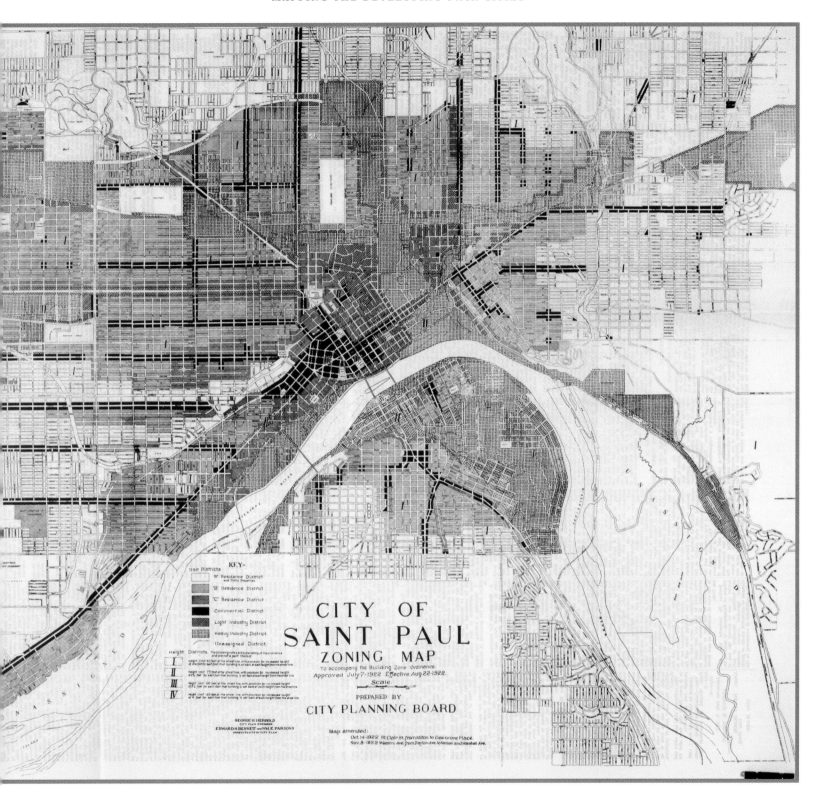

Preliminary City Plan,
from City Planning Board, Edward H. Bennett
and William E. Parsons, consultant city planners,
and George H. Herrold, city plan engineer,
Plan of St. Paul: The Capital City of Minnesota.

ST. PAUL: COMMISSIONER OF PUBLIC WORKS, 1922

17 X 23.5 INCHES

AUTHOR'S COLLECTION

by undesirable land uses, particularly commercial and industrial activities.

St. Paul's zoning map set out six forms of land use, including three residential densities ranging from single family (A) to multiple family (C). It also included one level of commercial use and two types of industrial districts. In reality, the planners followed existing land uses. Middle- and higher-income neighborhoods were well-protected; streets with streetcar tracks were classified commercial or as high-density housing. The areas abutting or adjacent to railroad tracks and the river were classified industrial. In the map detail (page 160), the riverbank settlement upstream from downtown known as Little Italy is zoned industrial, while affluent Summit Avenue is zoned A residential. They are separated by a transition zone of apartments and town houses on Crocus Hill and Ramsey Hill. Clearly the planners were conservative by today's standards in that they attempted to guide future development into easily recognized and acceptable patterns.

St. Paul's colored city plan (at right), which appears simpler than the zoning map, focuses on transportation and open space. The system of roads and railroads emblazoned here was essentially the existing pattern, with some important innovations. The most significant is the modern Shepard Road along the river, indicated on the map as the Forest Reserve Road, and the proposed alteration of the riverfront at the foot of the cliff below the central business district. The planners wanted to preserve most of the riverfront, other than the West Side River Flats, which were zoned for heavy industry and on this map were scheduled for an extensive barge harbor rather than an airport. The flats were eventually converted to an industrial park in the 1960s, but the city's major harbor stayed across the river on the downtown side. The proposed forest reserve around Pig's Eye Lake was only partially created. But the proposal for a large park along the river eventually was realized.

The plan's red areas are particularly interesting because they mark the city plan's call for the implementation of Cass Gilbert's ideas for a grand boulevard approach to the capitol that would begin on the bluff west of downtown. Unlike Gilbert's original ideas, however, this plan calls for the development of a large and stylish civic center just east of where the capitol boulevard intersects with the crest of the bluff. Although this government center did not materialize, for years the site was called the Civic Center. (It is now occupied by a cultural district, convention center, and hockey arena.)

These new functions were not part of the dreams of the planners of St. Paul in the 1920s, who seem to have looked forward to a continuation of the industrial city, improved with new parks, rationalized street patterns, and zoning. Three decades later, the building of freeways, slum clearance programs, elimination of streetcars, and booming suburbs would make their plan unworkable. Nonetheless, the zoning law they adopted protected most of the middle-class neighborhoods from commercial encroachment, and today St. Paul's greatest pride is its many and diverse neighborhoods.

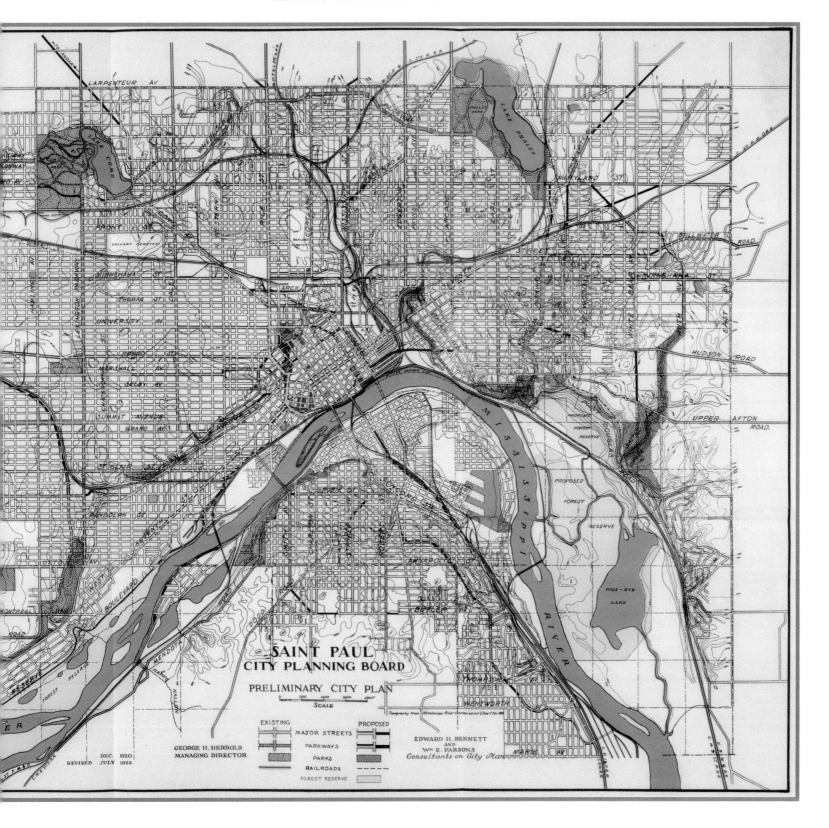

Natural Areas, Central Segment, St. Paul, 1935,
from Calvin Schmid, Social Saga of Two Cities:
An Ecological and Statistical Study of Social Trends
in Minneapolis and St. Paul.

MINNEAPOLIS: MINNEAPOLIS COUNCIL
OF SOCIAL AGENCIES, 1937

8.5 X 11 INCHES

AUTHOR'S COLLECTION

In 1937, the Minneapolis Council of Social Agencies published a report titled *Social Saga of Two Cities: An Ecological and Statistical Study of Social Trends in Minneapolis and St. Paul.* Its author, Calvin F. Schmid, a sociology professor at the University of Minnesota, produced this massive compendium of statistics, graphs, and charts based on research funded by grants from the Federal Writers' Projects. The grants employed a large but unknown number of young social scientists (nineteen are given special thanks) to gather and analyze data on population trends, housing, and "social and personal disorganization." The massive book has over 400 pages and 200 figures. A second volume of nearly 500 pages is titled *Guide to Studies of Social Conditions of the Twin Cities.* Together these encyclopedic works provide a detailed view of the city from the perspective of what was then called Human Ecology.

The Human Ecology approach focused on the city as a set of communities, called natural areas. According to this theory, urban communities are analogous to natural plant and

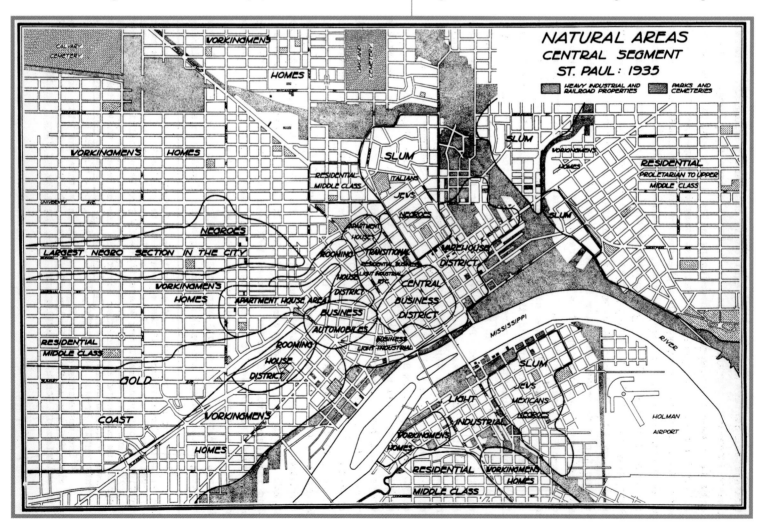

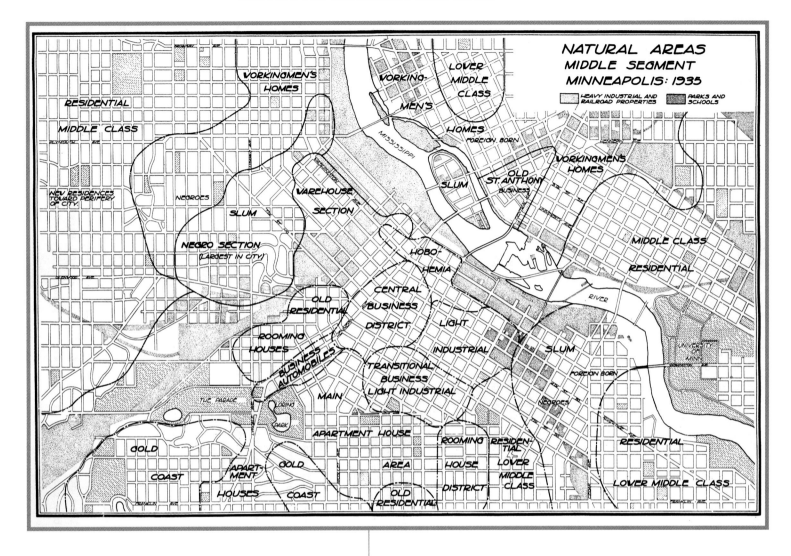

Natural Areas, Middle Segment, Minneapolis, 1935

animal communities studied by biological ecologists. This view of cities, which originated at the University of Chicago, sought ways to understand how individuals functioned in the large industrial cities. Human Ecologists also believed that societies trended toward equilibrium, and they therefore tried to understand criminal and other antisocial behaviors that seemed to upset equilibrium. By combining many variables, scholars could then map out the "natural" social areas and identify the regions where antisocial behaviors degraded communities.

Human Ecologists labeled these Twin Cities maps "Natural Areas" because they believed that there were natural laws governing societies comparable to the laws of nature. These laws worked outside human control and created natural areas of the city. The areas of greatest concern were slums, where people lived in degraded and dangerous conditions, and the social scientists mapped these areas block-by-block.

The Minneapolis map is more complicated that that of St. Paul because of the larger apartment-house districts and more rooming houses. At the western edge of Minneapolis's Hennepin Avenue Bridge is Hobohemia, a special sort of slum occupied by migratory workers. These workers rode railroad cars to work as field hands in the large commercial grain farms of the Red River Valley in the summer and in the winter worked as lumberjacks in the northern forests.

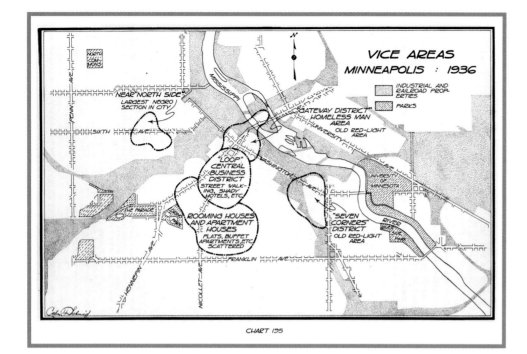

Vice Areas, Minneapolis, 1936

Social scientists and community leaders of the time were disgusted by the perceived growth of prostitution and venereal disease in the city. Convinced they could predict the presence of prostitution by examining the geography of the city, a Human Ecology research team thus confidently mapped Minneapolis's "natural" vice districts. This kind of mapping put the stigma on female prostitutes rather than on their customers, who lived in other residential areas that go unlabeled on the map.

Souvenir Guide Map of Saint Paul, Showing Places of Historic, Scenic and General Interest.

1953[?]

8.5 X 17 INCHES

RAMSEY COUNTY HISTORICAL SOCIETY

The *Souvenir Guide Map of Saint Paul* lets us know exactly what the movers and shakers of St. Paul in the 1950s thought made their city great. It was a "Family Sized Metropolis," "America's Friendliest City," and "Gateway to the Northwest." Furthermore it was a scenic and interesting place in which to drive around.

When this map was published, St. Paul had almost reached its population and industrial peak. On the East Side, 3M and Whirlpool were employing hundreds of blue-collar workers. The central business district had all the elements a downtown required. The First Bank Building, then the tallest in the area, dominated government centers, state and local offices, hotels, and corporate headquarters. Transportation still focused on railroads, and St. Paul was the center of railroading. Northwest Orient Airlines had established its headquarters in the Midway area, where it was joined by the "world's biggest trucking center." These two would soon combine to change the nature of long- and medium-distance travel and shipping in the United States and St. Paul. Also in the Midway, Gould Battery and heavy railroad mechanical works were still flourishing, and University Avenue was a center for a diverse range of industries, from printing to mattress making and paper manufacturing.

Across the river from downtown was American Hoist and

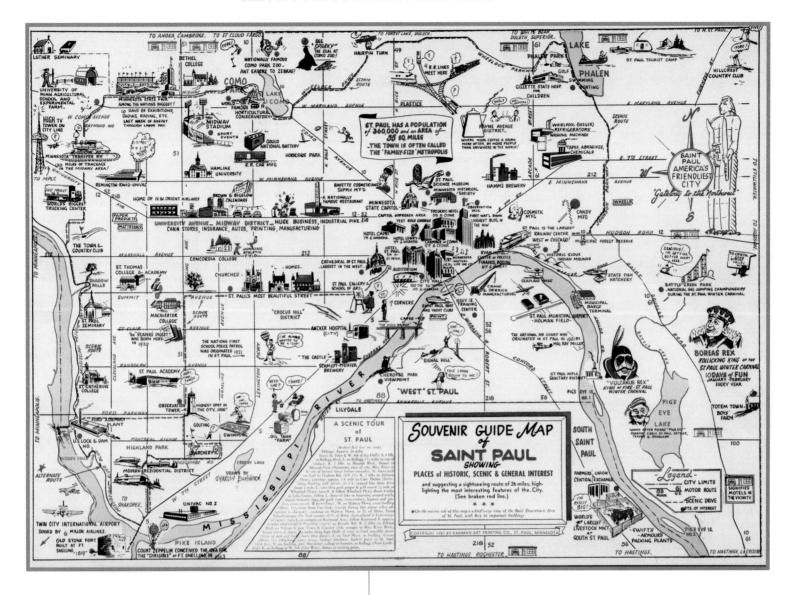

Derrick, the scene of manufacturing of heavy lifting equipment for export. The youthful computer industry was going through its first growth spurt, with Univac moving some of its operations out of the Midway to the bluff overlooking Fort Snelling. Evidence of the old St. Paul abounded—two breweries were still flourishing and the stockyards were listed as the world's largest. Just about everything a household needed, including cosmetics and candy, was still produced within the city's limits.

Older St. Paulites, whether snowbirds or year-round residents, look back at the place and time depicted on this map as St. Paul's golden era. Corporate leadership promoted civic festivals like the Winter Carnival with genuine enthusiasm. World War II was over and domestic life was looking up. St. Paulites had no trouble seeing their town quite independent of Minneapolis and other surrounding communities, and so there was no need to show them on this map.

Within the next fifty years, however, deindustrialization transformed the cityscape, culture, and economy of St. Paul. Most of the businesses featured so prominently on this map relocated or went out of business. St. Paul remains a popular place to live, but the city's 1950s landscape of heavy industry and downtown offices created by the railroad and streetcar has changed almost beyond recall.

Predominant Land Use 1958.

ST. PAUL: TWIN CITIES METROPOLITAN PLANNING COMMISSION, 1959[?]

22.3 X 17.5 INCHES

AUTHOR'S COLLECTION

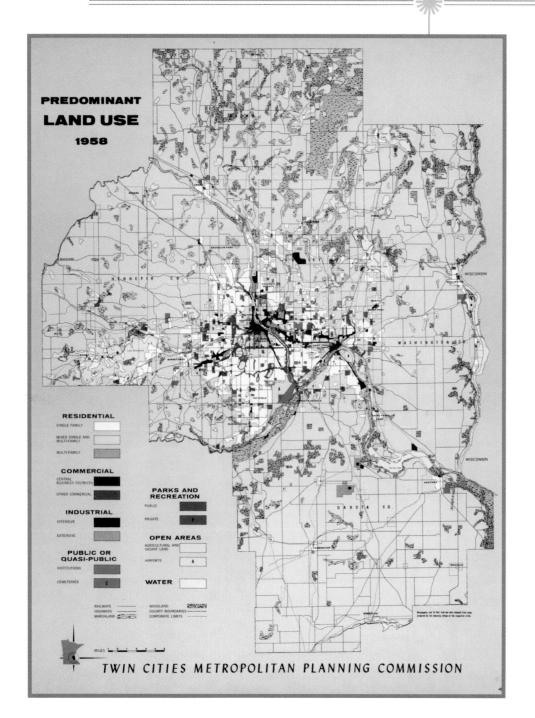

During the mid-twentieth century, the Twin Cities metropolitan area grew in wealth, population, and area. By 1957, the rapid rate of expansion had sparked a wave of incorporations of suburban communities until the metropolitan area included 365 municipalities. The patchwork of governments, some very large, others minuscule, struggled to cope with emerging landscape issues. Traffic congestion increased, and problems of water quality and waste disposal emerged. Some communities attracted new commercial development and enjoyed expanding tax revenues; others languished as tax-poor bedroom communities.

In response, local government, business, and civic leaders of varying political persuasions cooperated to create the Twin Cities Metropolitan Planning Commission (succeeded by the Met Council). Postwar urban planners were eager to create a new urban region. The massive conversion of farmland to new suburban communities gave them the wonderful opportunity to build the city anew, and this time to build it correctly, according to the best modern ideas and not just as a money-making machine for the real estate industry.

Work began with an inventory of existing land use. The Metropolitan Planning Commission published two maps that clearly illustrate the dramatic rate of urban growth experienced by the metro area. *Predominant Land Use 1958* depicts the impact of the explosive urban growth that began in 1946, the result of pent-up demand from lack of building during the Depression and the war years and expanding incomes. Federal government programs such as home loans for war veterans and home loan insurance made financing attainable for a huge segment of the labor force.

Perhaps most important, the automobile industry began creating attractive and affordable cars, and the great American love affair with the automobile was in full bloom. Workers could live in new housing away from the streetcar lines. In 1920, only 250 square miles of land lay within one hour of travel from the nearest downtown; with the advent of the automobile, that one-hour zone of easy access increased to 2,000 square miles.

Opportunities in housing construction drew large-scale developers to the Twin Cities for the first time. By 1958, single-family houses filled all of the partially developed outskirts of the core cities and first-ring suburbs and pushed into the country. These developments were based on non-freeway highway access or early freeway development, and most were primarily extensions of nearby central-city neighborhoods. Commercial areas were developed that expanded narrow streetcar strips to accommodate several wide lanes for automobiles.

However desirable these developments were, important problems accompanied this type and pace of growth. Main highways were few (and in poor condition), and only thirty-six arterial highways reached ten miles outside cities. Residential growth ran ahead of services and employment in suburbs. At one time, as many as 300,000 people used septic tanks, and an unknown number were still using even older techniques for disposing of waste.

This map shows the Twin Cities' early suburbs as filled in: Richfield, Hopkins, St. Louis Park, Golden Valley, Columbia Heights, Roseville, West St. Paul, and South St. Paul. The second ring suburbs of Bloomington, Edina, and Minnetonka would follow very soon. Most of northern Dakota County, Washington County, large portions of northern Ramsey County, and the majority of Hennepin County are still classed as agricultural or vacant land. The downtown cores, the streetcar commercial strips, and the railroad-oriented intensive industrial areas are strong and dominate the landscape. There are no freeways.

Twin Cities Metropolitan Area Generalized Land Use 1968.

ST. PAUL: METROPOLITAN COUNCIL OF THE TWIN CITIES AREA, 1968

22 X 28.75 INCHES

AUTHOR'S COLLECTION

This 1968 map looks to the future. Depicting the middle postwar urban expansion of the Twin Cities, it shows the agreed-upon route of "interstate freeways," construction of which was well under way. The plan called for a beltway encompassing the area that, in places, extended beyond the built-up zone. Interstate 94 ran east-west and Interstate 35 ran north-south, splitting to give both downtown Minneapolis and downtown St. Paul a freeway intersection.

In the decades served by this map, the outer city was becoming the focus for the economy of the metro area, as well as for the several state regional economies. Large corporations were moving operations to the suburbs, for example, 3M to Maplewood and General Mills to Golden Valley. The south Minneapolis Bloomington Strip was attracting a diverse set of commercial enterprises. An entire range of new development patterns emerged, including shopping centers and shopping malls, suburban office buildings, and entire corporate campuses. Industrial parks were developed for extensive and freeway-oriented operations. Older, heavy, railroad-based sites were vacated. The improved highways and freeways enabled residential developers to go farther afield, where they turned farms into subdivisions with larger lot sizes and the new cul de sac designs. Urban services follow the development of residential areas, and there was a notable boom in school building.

The southwestern sector of the circumferential Interstate 494/694 was built first in response to the rapid growth in that sector. Once complete, the freeway attracted even more growth. Interstate 35W, running south through Minneapolis, Richfield, and Bloomington, and on to Burnsville, was intended to improve the commute to and from downtown and create a new Main Street for the metro area. Its twin, Inter-

MINNESOTA ON THE MAP

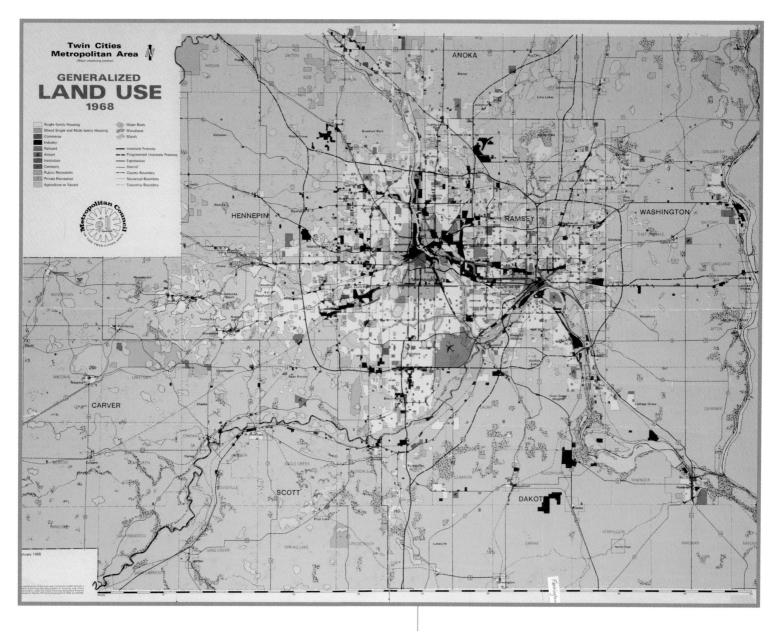

state 35E, intended to serve the east metro, was built later, which delayed growth in northern Dakota and Washington counties for several decades.

The decade between 1958 and 1968 was a critical period for the growing urban area. The older suburbs were filled, and areas like Brooklyn Center, New Hope, Medicine Lake, and Cottage Grove changed from rural to urban. The highway to Anoka became a growth zone. The changes on the edges were prompting changes in the center, and new social issues arose to challenge planners and leaders. There were two responses: the one at the regional scale involved a plan for the metro area. The other, at the city scale, focused on the revitalization of the center.

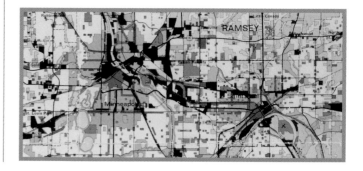

170

Constellation Cities, 1985,
from Report Number 5, 1968.

ST. PAUL: JOINT PROGRAM,
AN INTERAGENCY LAND USE–
TRANSPORTATION PLANNING
PROGRAM FOR THE TWIN CITIES
METROPOLITAN AREA, 1968

22 X 17 INCHES

AUTHOR'S COLLECTION

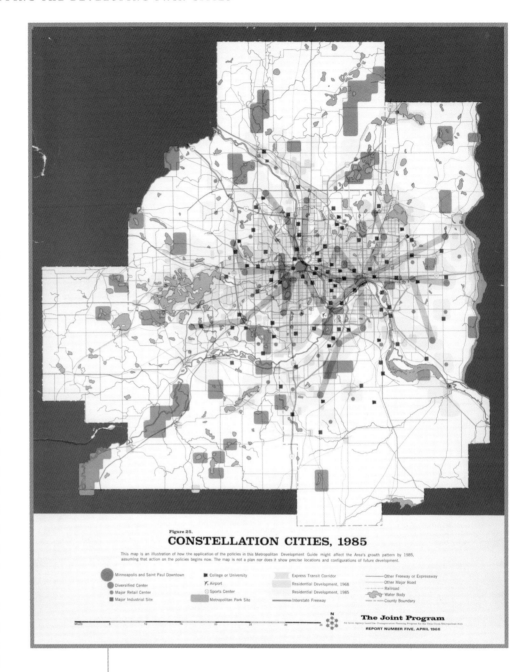

In 1964, professional planners at the Twin Cities Metropolitan Planning Commission published a report with the buoyant title *4,000,000 by 2000! Preliminary Proposals for Guiding Change.* Fervently believing that great changes were coming, they wrote, "It is the nature of a metropolis to change. A metropolis continually grows or declines, ages or renews, expands or contracts, speeds up or slows down. The Twin Cities is no exception." Given these perceived options, planning for growth was seen as the only rational choice. (Planners misjudged the area's long-term growth rate; by 2006, the population had reached only 2.81 million.)

In the early 1950s, the Dayton Company had boldly banked on the future by hiring architect Victor Gruen to design for the suburb of Edina the country's first enclosed and air-conditioned shopping mall. Southdale, as it was called, was to be the nucleus of a suburban community. Freeways were under construction, and the bucolic landscape surrounding the cities was giving way to subdivisions and industrial parks. Plans were needed. The commission wrote: "Indecision or leaving the area's growth to chance won't preserve what we value for long. We would stand to lose much of what we cherish about the area and gain little in return. This is no way to create the best possible living environment for our later years or for the productive years of our children."

Metropolitan planning commissioners and their staff soon realized they lacked power to enforce planning, however, and in 1967, in what today seems an amazing gesture of civic forward thinking by the region's many small political entities, a new Metropolitan Council replaced the planning commission. It has more power to make municipalities conform to regional goals and to regulate the expansion of sewers and transportation facilities.

Twin Cities, 2000 AD: Alternative Patterns,
from *Report Number 4,* January 1967.

ST. PAUL: AN INTERAGENCY
LAND USE–TRANSPORTATION PLANNING PROGRAM
FOR THE TWIN CITIES METROPOLITAN AREA, 1967

11 X 8.5 INCHES (EACH)

AUTHOR'S COLLECTION

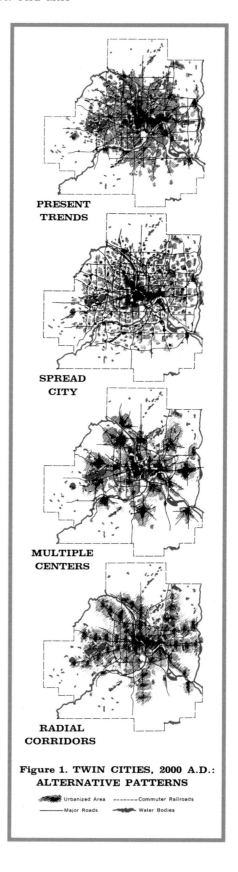

Figure 1. TWIN CITIES, 2000 A.D.: ALTERNATIVE PATTERNS

In 1968, the Metropolitan Council published its plan in *Twin Cities Area Metropolitan Development Guide,* which provided the basis for subsequent development guides to manage growth in the Twin Cities. The plan grew out of widespread discussion of a set of possible futures for the area: "Present Trends," "Spread City," "Multiple Centers or Constellation Cities," and "Radial Corridors." In the end, this Constellation Cities model was accepted because it seemed to have something for everyone. In the words of the planners, the Constellation Cities pattern offered the best approach to the goals of the community by expanding the downtowns; building major outlying shopping and office centers, industrial parks, extensive recreational spaces, and effective transit; and encouraging community formation.

It is impossible to know if the Twin Cities would have developed into today's sprawling, multicentered metropolitan region if the Constellation Cities plan had not been used as a guide. In many respects, the plan did reflect current trends in the land market, although the transit plans were not realized. It is clear that the acquisition of large regional parks and the controls on urban expansion were possible because community leaders and developers could use these maps to envision the future for the entire area and understand how their efforts in local subdivisions could fit into a much larger urban system. We are accustomed to seeing maps of the metro region today, but these maps were revolutionary in the 1960s, and they made the Minneapolis–St. Paul metropolitan region different from similar-sized urban areas around the country.

MAPPING THE DEVELOPING TWIN CITIES

Metro Center '85 Illustrative Site Plan, from *Metro Center '85 Study for the Development of Program and Priorities for Expanded Job and Investment Opportunities in Central Minneapolis.*

MINNEAPOLIS: PLANNING AND DEVELOPMENT, 1970

11 X 11 INCHES

AUTHOR'S COLLECTION

Fearing the impact of suburban growth and innovations such as evidenced in Victor Gruen's Southdale, the Minneapolis Downtown Council determined to rebuild and revitalize its downtown. The city hired Larry Irvine to head the city's planning department, and under his guidance the department and consultants produced three significant planning documents: Central Minneapolis Plan in 1959 (never adopted), *Metro Center '85* (1968), and *Minneapolis Metro Center, Plan . . . 1990* (1978). The plans were based on five principles: 1) downtown Minneapolis should remain the most important place in the Upper Midwest; 2) the city center should be compact; 3) accessibility to the center should be improved; 4) downtown should be a comfortable and friendly place for people; and 5) development should result from a partnership between government and the private sector.

Although never formally adopted, the 1959 plan was critical in the development of the city. Unabashedly optimistic in the face of suburban growth and high office vacancy rates in downtown, it boldly set out to counter economic stagnation in the city center and contend with the standing stock of older buildings. Specifically, the plan called for 1) constructing a ring road around downtown; 2) building a freeway parking terminal; 3) promoting new high-rise buildings; 4) finishing blight clearance and renewal efforts in the gateway district; 5) developing a park on Nicollet Island; and 6) promoting Nicollet Mall.

Nicollet Mall was the first of its kind in a large North American city, although pedestrian-only streets were common in redeveloped European cities. The building of the mall received the highest priority because it was seen as key to boosting retail trade and improving the market for office space. Most planners and business leaders believed it would work because the shopping area was compact and busy, and it contained four-fifths of the city's retail space. Work began in 1963 and was finished five years later. It has become one of the iconographic landscapes of Minneapolis, copied by cities large and

Riverfront Housing and Cultural Center Illustrative Site Plan,
from *Metro Center '85 Study for the Development of Program and Priorities for Expanded Job and Investment Opportunities in Central Minneapolis.*

MINNEAPOLIS: PLANNING AND DEVELOPMENT, 1970

5.5 X 11 INCHES

AUTHOR'S COLLECTION

small all around the world. In the 1970s, the mall became a symbol of the new life available to professional women when one of the decade's most popular weekly TV shows began with the bright, talented, and vivacious Mary Tyler Moore pirouetting and tossing her hat in the air amid winter shoppers on Nicollet Mall.

In 1970, the *Metro Center '85* plan was issued as an elaborately illustrated $15 hardcover volume. This product of the innovative and socially conscious 1960s presented a mod-

ernistic view of an exciting, upbeat city filled with new governmental programs and urban culture. Its authors recognized the continued retail flight to suburban malls but believed that Nicollet Mall would hold enough activity to maintain the central business district. Improved transportation, clustered functions in core districts, and diverse races and lifestyles were to be celebrated. In addition, visual design was to be emphasized. Philip Johnson's recent dazzling IDS Center enabled the city to boast one of the finest skyscrapers of the modern school of design, and the plan enabled a new burst of downtown building that produced several excellent buildings in postmodern design. By the end of the century, Minneapolis had one of the continent's most beautiful downtown skylines.

Every plan needs a bold vision. For *Metro Center '85* it was a 25-acre cultural center that was to extend from the site of the old Milwaukee Station to the river. The development called for the removal of the obsolete and vacant mills that lined the west bank of the river near the Falls of St. Anthony and the lock and dam system of the Mississippi. The plan included a symphony hall, two theaters for stage productions, an art museum, and a huge exhibition hall for a museum of technology, as well as grand parks, promenades, and a marina. There were to be several outdoor spaces for people to congregate, view sculpture, and eat. The planners did not establish a schedule for these changes, however, perhaps because they understood how difficult they would be to implement.

Shortly after the plan was published, the board of the Minnesota Orchestra decided not to wait for the plan's lengthy approval process and instead built a new performance hall at Nicollet Avenue and Twelfth Street. The riverfront development was thwarted by complex property holdings. The Science Museum of Minnesota remained in St. Paul, where it eventually built a new facility on the Mississippi. In time, a museum was established on the site identified in the plan, but it was not a sleek modern structure built to glorify technology and science but a refurbished and partially restored mill ruin.

This era's planners did not anticipate the building of the Metrodome, the Loring Greenway, or the rapid deindustrialization of the core city, but their vision, general goals, and guidelines for development were appropriate. The 1970 plan is a bold statement of modern planning and architecture theory published at the end of that era before the dawning of postmodern architecture and planning. Like the earlier City Beautiful plan of 1917, *Metro '85* presented the latest ideas about urban life but failed to control the development process.

9

LANDSCAPES OF RECREATION

Today's Minnesotans know how to enjoy their state in all its seasons. Seeing the state as a source of unlimited opportunities for fun, we manage its landscapes for recreation and entertainment and to foster cultural tourism, or the conscious appreciation and visiting of those resources.

A century and a half ago, Minnesota was viewed primarily as a landscape of production, a place where hard work would result in a better life. Farming, logging, mining, and related manufacturing dominated the economy and culture of the state. Although these economic activities and landscape still exist, deindustrialization has brought change in many urban areas. Added to this mix of productive activities is the economy of consumption and landscapes of consumption.

Minnesotans have also created signature or iconic landscapes that are marketed by the state government and the tourist and recreation industries. The state's signature landscape relates to its 10,000 lakes. Visions of canoeing in pristine lakes and streams, fishing on clear blue lakes, or hunting in dense pine or hardwood forests are conjured up by the name Minnesota. These summer dreams are complemented by those of winter activities like snowmobiling, ice fishing, and cross-country skiing. The emotional "landscapes" that facilitate these activities draw thousands of people to the state, but so many newcomers threatens the very features that drew them here. Consequently, a large management effort is needed to maintain the state's attractiveness.

Minnesota is one of the very few states where a person can travel from one of the nation's largest metropolitan areas into roadless wilderness, complete with a growing population of timber wolves, moose, and deer, in about five hours. Heading north from the Twin Cities to the wilderness, a traveler will pass through farmland, small towns, resort developments, second-growth forests, and unoccupied wild land. Travelers going other directions see the rich cultural legacy of the transformation of the land and smaller urban centers that portray their unique features and special cultural traits through museums, preserved buildings, and an incredible range of local festivals. Some of the festivals feature local products like corn; others feature events like tractor pulls that hark back to a previous era. All the communities have golf courses, athletic fields, and parks that provide settings for the consumption of culture and place.

Langwith's Pictorial Minnesota, The Land of 10,000 Lakes.

ST. PAUL: MCGILL-WARNER, 1921

AUTHOR'S COLLECTION

This transformation of places from landscapes of production to landscapes of consumption was made possible by the growing wealth of the American population, the expansion of leisure time, and improvements in transportation, especially auto and air travel. All the covers of the state maps produced by the 10,000 Lakes Association epitomize this cultural shift.

Minnesota Invites You to Live, Work, Play in the Playground of 10,000 Lakes.

ST. PAUL: MINNESOTA DEPARTMENT OF CONSERVATION AND MINNESOTA TOURIST BUREAU, C. 1933

31 X 21 INCHES

LAURA KIGIN COLLECTION

One of the earliest efforts by the state to promote automobile tourism in Minnesota may be seen in this Department of Conservation map produced during the terms of Governor Floyd B. Olson (1931–36). It includes a "letter" from the governor that begins: "To Everybody Seeking Happiness—We have a good time for all of you in Minnesota, whatever your inclination." The small inset map showing major highways connecting Chicago and Minnesota suggests that the map may have been produced for distribution among the sixty-four million visitors who attended the Chicago World's Fair of 1933–34, which celebrated "The Century of Progress." Railroad companies had promoted tourism in an earlier era, their efforts focusing on destinations sited along the tracks, but the makers of this map did their best to promote features all over the state. Governor Olson was an ardent supporter of programs to help Minnesota to weather the Depression, and tourism by middle-class families held promise. No great investments of state money were needed. The lakes, fish, forests, and game animals were already in place. All that was needed was a marketing campaign.

The focus of the *Live, Work, Play* map is broader than tourism, however. The symbols and text combine to create an image of economic potential for both investors and workers. The map identifies several places—a packinghouse in Austin, stockyards in South St. Paul, and vegetable canneries in Le Sueur—that could hardly be considered tourist destinations. The symbols and roads are designed to provide the viewer with the impression that the new state highway network will make it easy to get around the state and that there is something interesting to do or see in all parts. Where points of interest are few, the mapmaker filled in the blank spots with generic information. Therefore, the southwestern part of the state contains several nonlocal symbols and facts, such as the number of airports in the state, railroad mileage, and production figures for various crops. On the other hand, the hilly southeastern portion of the state is dubbed Little Switzerland and includes only a handful of specific place labels. The absence of any corporate names indicates the lack of

"Minnesota, Where Dreams Still Come True," cover of Minnesota Invites You to Live, Work, Play in the Playground of 10,000 Lakes.

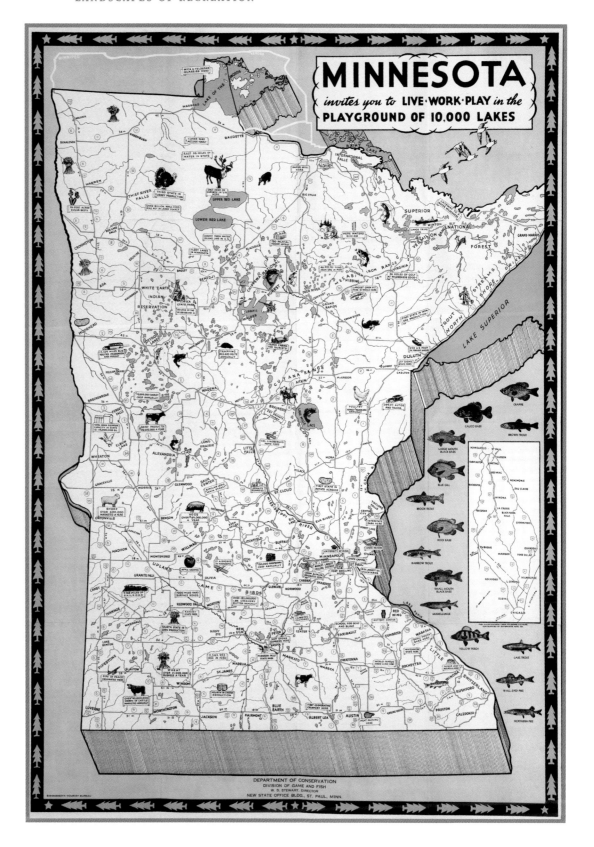

business sponsors for the map. This is especially noticeable in the Brainerd area, where one small box mentions 297 nearby resorts, and farther to the northeast, where another indicates the Gunflint Trail as the gateway to a water wilderness. It is significant that the Gunflint Trail and the unnumbered road to Ely through the Superior National Forest are designated with a line the same color and width as the primary state highways.

This map spawned a huge number of imitations, updates, and improvements in the succeeding seventy-five years. The state continues to distribute free highway maps as a means of promoting tourism. Some of the messages from this early map continue today, especially the themes of outdoor recreation and a landscape for consumption. The message remains the same: Minnesota is a place where dreams come true.

Langwith's Pictorial Map of Minnesota— Come-to-Minnesota Club.

MINNEAPOLIS: LANGWITH MAP CO., 1920S

37.75 X 24 INCHES

AUTHOR'S COLLECTION

In the early 1930s, the Langwith publishing firm of Minneapolis issued one of the most complicated pictorial, or graphic, maps of Minnesota ever produced. The company survived in the highway map business for about a decade after first producing a road map of the United States that folded up and a highway map for Pure Oil and perhaps other petroleum companies. This time its client was the "Come to Minnesota Club," whose members included establishments in the hotel, resort, and travel industry, including several gasoline retailers. This innovative marketing plan was to involve quarterly updates of the map and its directory; most likely the only thing updated was the list of commercial establishments waiting to help the tourists. Consumers who purchased this map could visit participating gas stations and redeem the old map for a new one, for the price of $2.50 up front. The variety of schemes developed to sell maps of Minnesota was truly extraordinary, although this one apparently failed.

The colorful poster-sized map bristles with information, stereotypes, and popular lore of the time. The covers promise it all: Agriculture, History, Industry, Education, Recreation, and Geography. The designer was clearly of the "too much is not enough" school of cartography. Lakes, roads, towns, and cartoons are all crowded onto the map. The tip of the Arrowhead Region has been nipped off to fit the page better, and the Northwest Angle has been given back to Canada. The hodgepodge of information does indeed include all sorts of data about the production of commodities, famous people, and individuals or couples leading wonderful lives of leisure. Although the map shows the location of roads and towns with reasonable accuracy, the history it tells is farcical. Apparently the cartoonist confused Lieutenant Zebulon Pike with Colonel George Custer, because we are informed that it was Custer who negotiated a treaty with the Native Americans.

The images of tourists depicted on the map include sophisticated travelers riding horses, lounging, reading books, cooking over an open fire, fishing, hunting, and walking. There is no doubt that all these activities were a part of the recreation business of the time, but costumes like these must have been exceedingly rare. Extra tidbits include a man skinny-dipping in Lake Superior and Dr. Mayo's observation that the Rochester area has many beautiful drives that are unmatched anywhere else in the world. But the best is the claim that Iowa's Lake Okoboji is "Minnesota's last crystal Clear Blue water lake."

Credit for the design of this map goes to someone who signed it simply "Henri" and depicted a red-nosed painter before an easel. The map may have done Minnesota's tourist industry more harm than good. ✻

Alexandria Lake Region in Minnesota.

VIRGINIA, MINN.: FISHER MAP CO. AND ALEXANDRIA CHAMBER OF COMMERCE, 1939

17.6 × 22.4 INCHES

MINNESOTA HISTORICAL SOCIETY

Working with the Fisher Map Company of Virginia, Minnesota, between 1939 and 1955, Frank Antoncich produced a handful of cartoon or pictographic maps of popular vacation areas in northern Minnesota. This 1939 map of the Alexandria Lakes region northwest of the Twin Cities is representative of his work and the genre. The overall purpose of the map was to help actual and potential tourists build a mental image of the area and a sense of the relative location of key places and features in the environment and the area's history. Though not designed as a navigation map, it nevertheless depicts the area's road network and could be used for jaunts, if not long trips. The map prominently features the intersection of two transcontinental routes, the Winnipeg to New Orleans route, State Highway 29, and U.S. 52, the east-west route linking New York City and Seattle.

Unlike the images on the official highway map of the state, Antoncich's illustrations are filled with homespun humor. Antoncich (or his editor) was something of a punster, encouraging visitors to come to Alexandria "for the rest of their lives." One of the few popular maps with illustrations of hunting, it depicts with stereotypical caricatures the "Chinese" pheasant population and the area as a productive breeding ground for ducks. Antoncich also notes the increasing mechanization of agriculture, with anthropomorphic horses engaged in labor disputes and watching a tractor taking over their place in the economy. The illustrations also poke fun at tourists, noting, for instance, the changing styles of swimwear brought to the northern lakes by the fashionable travelers. No map of this genre depicting the Alexandria area would be complete without some reference to the Kensington Rune Stone, and Antoncich provides his readers with horn-helmeted explorers and their stone, which probably means he doubted the artifact's authenticity, as did most non-Scandinavians.

This illustrated map is part of a larger genre that includes popular postcard maps. The Fisher Publishing Company also produced popular postcard maps with humorous figures. ✻

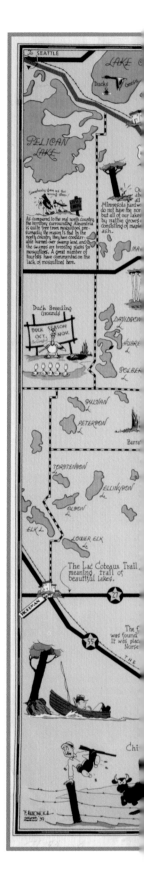

LANDSCAPES OF RECREATION

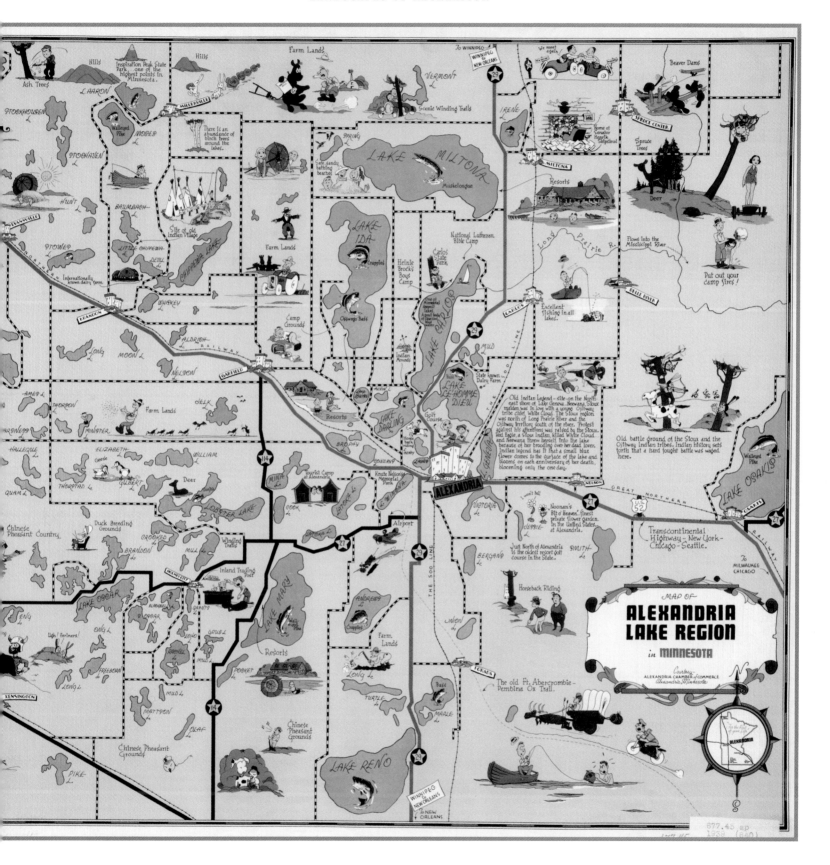

Itasca County, Minnesota: The Heart of the 10,000 Lakes Country.

ST. PAUL: MCGILL-WARNER CO., 1923

18 X 16 INCHES

LAURA KIGIN COLLECTION

McGill-Warner Company of St. Paul printed this handsome map of Itasca County in 1919 and again in 1923 for an unidentified organization promoting tourist attractions within the county. The organization clearly understood the significance of roads to the new tourist industry, as travelers came in search of natural sites such as the source of the famous Mississippi River. With detailed maps like these, sportsmen and families could wend their way along the narrow highways and country roads into the woods and find the perfect lakeshore cabin, resort, or campsite. If they had car trouble, they could see where the nearest farmstead was with a team of horses to pull the car out of a mud hole or tow them into town.

Because the map was published before the state highway numbering system had been established, major paved roads are named and identified as automobile roads to differentiate them from the unpaved "country roads." Even the smallest settlements are identified, along with newly established resorts and fishing camps. On the back of the map, the area's abundant fish and game are emphasized as part of the year-round sporting activities in the northern forest. Not mentioned are golf, tennis, or anything associated with urban recreational activities. Special attention is given to the wonders of the mining operations on the Mesabi Iron Range, and it is proudly noted that more earth was moved in digging these mines than in the digging of the Panama Canal.

This map is significant because it portrays the landscape of consumption emerging from the preexisting landscape of production based on iron-ore mining, timber, and subsistence agriculture. Community promoters sensed that they would be able to convert mining into a tourist attraction but avoided discussing timber or the hardscrabble farms created on northern timbered lands. The county had taken the beautiful natural landscape and added to it the roads necessary for tourism. The map caption asks, "Can a lover of out-door sports ask for more?"

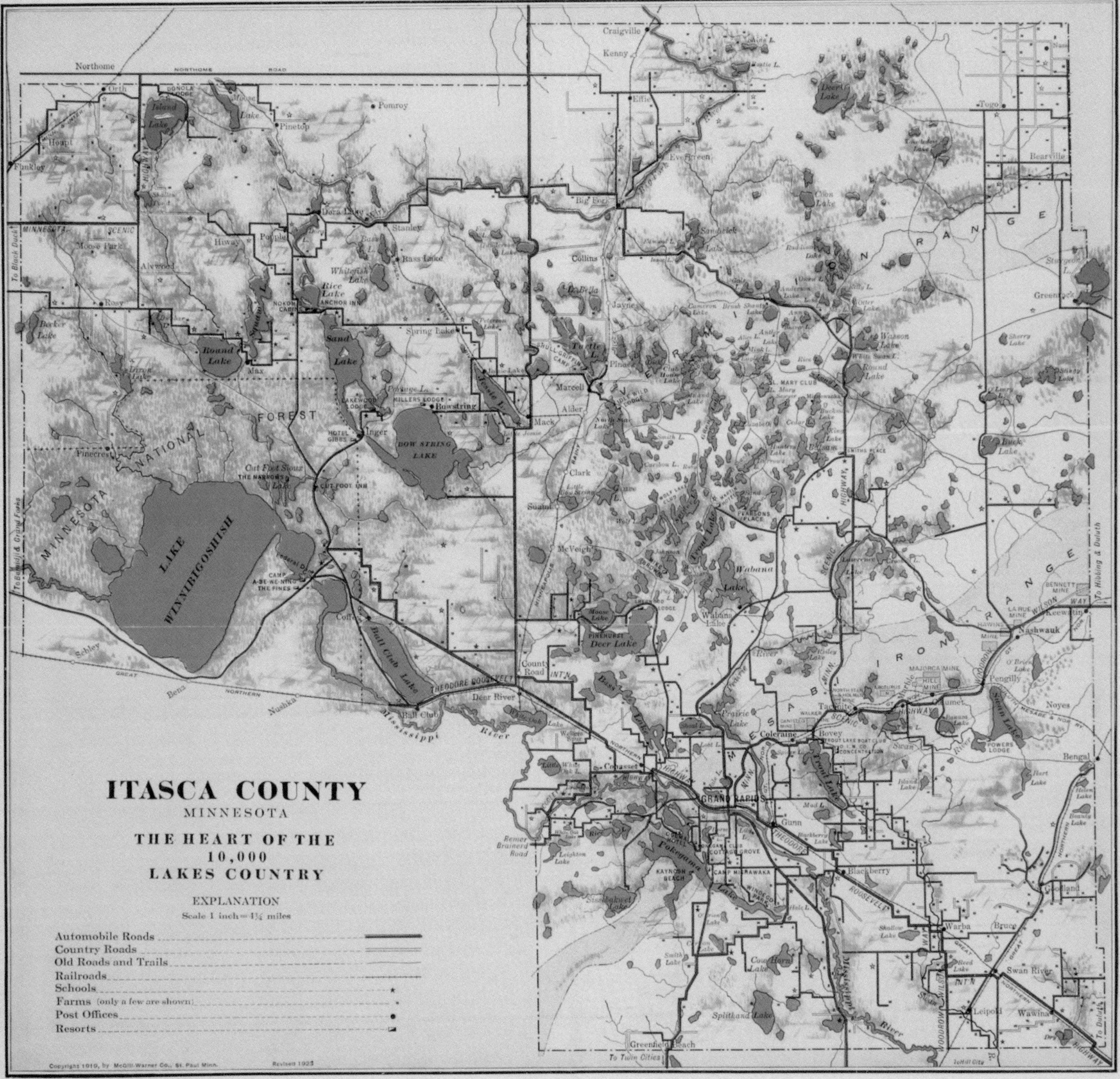

ITASCA COUNTY
MINNESOTA

THE HEART OF THE 10,000 LAKES COUNTRY

EXPLANATION
Scale 1 inch = 4¼ miles

- Automobile Roads
- Country Roads
- Old Roads and Trails
- Railroads
- Schools
- Farms (only a few are shown)
- Post Offices
- Resorts

Copyright 1919, by McGill-Warner Co., St. Paul Minn. Revised 1923

Countless, beautiful lakes and rivers, game aplenty, excellent roads.
Can a lover of out-door sports ask for more?

MINNESOTA ON THE MAP

Selected Minnesota postcards.

LAURA KIGIN COLLECTION,
AUTHOR'S COLLECTION, AND
LURED TO THE LAKE

ostcard maps quickly convey a hand-sized image of the state that combines some of the accuracy of a map with a wide variety of tourist-oriented images and icons. These appealing, often humorous, maps usually include the main towns and cities but not always the highway system. Any traveler who tried to use the image as a road map would need to watch out for a range of giant creatures, including Paul Bunyan and his blue ox. Other frequently appearing images include Indians, often in inaccurate dress, nat-

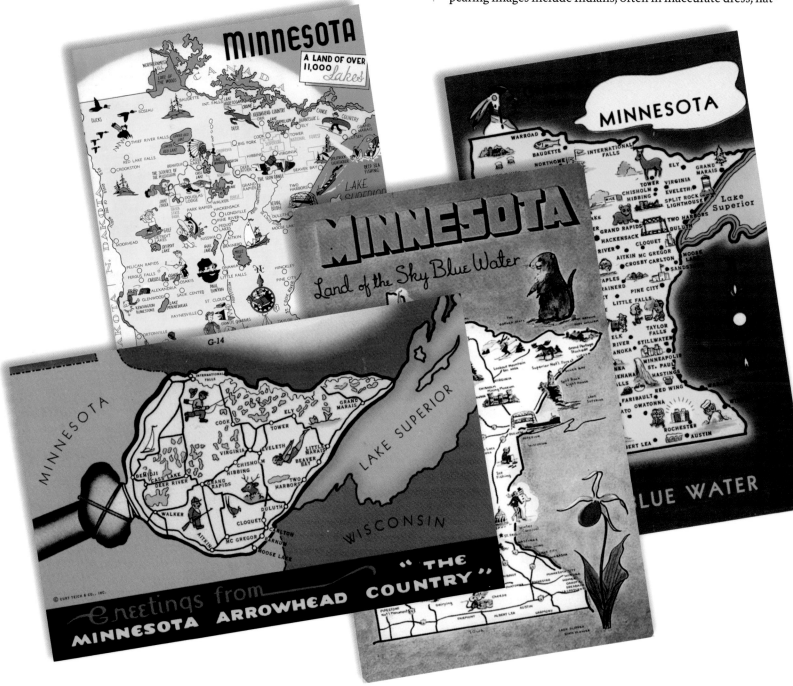

ural sites such as Minnehaha Falls, landmarks such as Split Rock Lighthouse, and happy people fishing, golfing, skiing, swimming, canoeing, boating, and camping. Gophers and lady's slipper flowers abound.

Map publisher Langwith even issued a miniaturized version of its *Come to Minnesota Map* as a postcard, which shows the mileage between major points of interest. Its detail and broad humor are unmatched by any other postcard publisher.

Postcard maps were also printed for major recreation areas such as the northeastern Arrowhead Region.

Postcard maps rendered in bright colors are especially eye-catching. They signaled to anyone receiving them that the senders were on vacation or traveling for fun. Handwritten messages on the other side of the card were appropriately short, signaling that the sender was having too much fun to bother writing at length.

Minnesota State Parks, State Forests, and Recreational Areas.

ST. PAUL: MINNESOTA DEPARTMENT OF CONSERVATION, 1941

30 X 21 INCHES

MINNESOTA HISTORICAL SOCIETY

This map of Minnesota's state parks, forests, wayside rests, monuments, and recreation areas, produced by the state Department of Conservation in 1941, follows in the tradition of the department's 1931 map, but it uses more prominent illustrations to attract the reader's attention and promote the state's recreational amenities. Again, each part of the state gets some attention to avoid the impression of favoritism. Because the state's tourist industry was focusing on a lakes and woods motif, southern and western Minnesota had few landscapes to advertise, but rough wooden frames democratically highlight little-known places that viewers could be excused for not knowing about.

Governor Harold Stassen and officials of the Conservation Department had reason to be proud of Minnesota's parks. The state was the second in the nation to create a park system when it established Itasca State Park in 1891. When this map was drawn, the department could boast about the ten new parks that had been added to the system in 1937. Around the state, the Works Progress Administration had improved major scenic overlooks with limestone and boulder walls and bridges. New benches lined new trails, and picnic facilities and park buildings had been improved. The administration's goal for this map was to have citizens take ownership of the park and forest system, as well as to attract visitors who would spend money and give the economy a boost. It was hoped that tourists would not fit the apocryphal description of the person who arrived on a two-week vacation in Minnesota with a new pair of bib overalls and a new twenty-dollar bill in a pocket—and changed neither.

Minnesota's park system continued to expand in the twentieth century in response to the growing number of auto-traveling tourists who had an interest in nature and wanted to explore the state's regions. Over the years, Minnesota's park system expanded to include sixty-six state parks and six recreation areas. The Department of Natural Resources, which is responsible for the parks, also manages eight waysides, a trail, and over fifty campgrounds in state forests.

Minnesotans have internalized their state's cultural map and developed a great appreciation for parks. Well over eight million people visit a park each year. Most are day-trippers, but nearly a million people camp overnight in parks. Out-of-state tourists use the parks, but 84 percent of state park visitors reside in Minnesota.

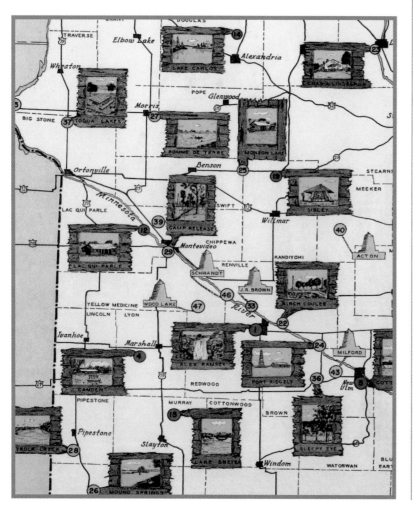

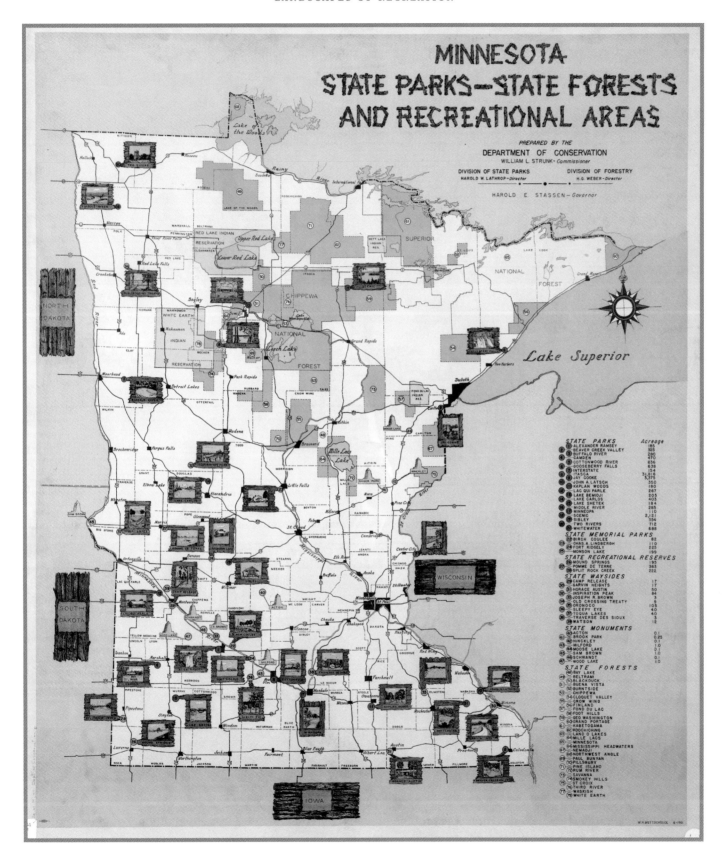

Map of the Minnesota Arrowhead Country.

ST. PAUL: MINNESOTA ARROWHEAD ASSOCIATION, 1941

16 X 26.5 INCHES

AUTHOR'S COLLECTION

The Boundary Waters Canoe Area Wilderness is Minnesota's unique recreational landscape. Not everyone in the state has canoed its solitary lakes and rivers but everyone knows it is there, "up north," somewhere. Comprising about one million acres, it has over 1,200 miles of canoe routes and portages. Part of Superior National Forest, the BWCAW was established to provide a place for wilderness and recreational activities, to protect threatened and endangered species, and to manage an important watershed. Each year about 200,000 people visit the BWCAW—a number that is limited to maintain the wilderness quality of the area—making it the most heavily used wilderness area in the United States. Many canoeists view Quetico Provincial Park across the Canadian border as an integral part of this wilderness.

The Superior National Forest was established in the early years of the twentieth century in response to popular pressure. Following World War I, auto tourism in the national forest grew markedly, partly in response to the promotional efforts of the Arrowhead Association, a group that published accurate maps of the region. Soon tourists and local timber companies were disagreeing on the best way to manage the

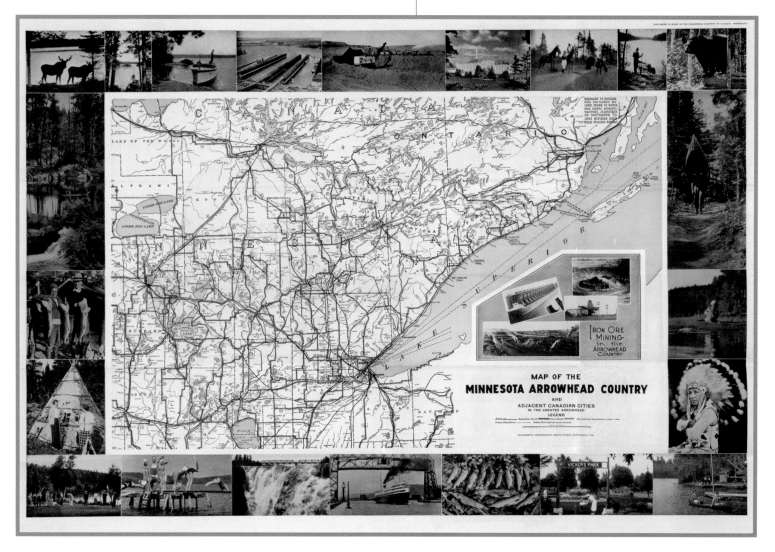

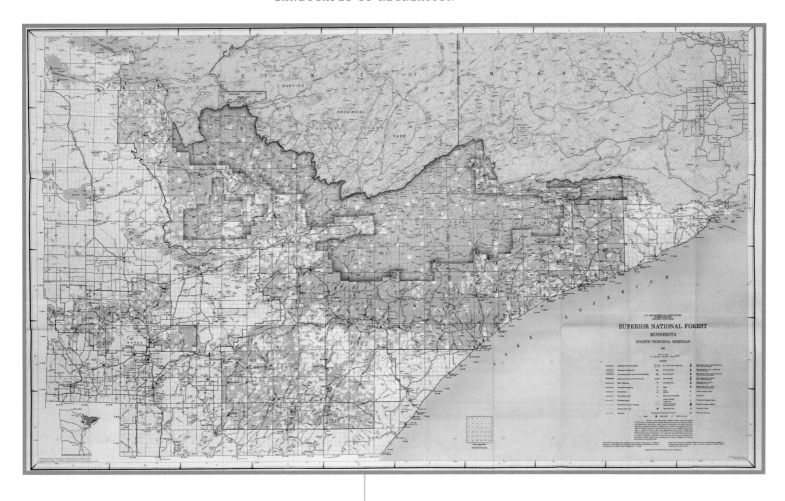

Superior National Forest.

DULUTH: U.S. DEPARTMENT OF AGRICULTURE, FOREST SERVICE, 1969

26.5 X 48 INCHES

AUTHOR'S COLLECTION

national forest, and between 1920 and 1940 a sharp political battle developed between people seeking to preserve the area's wilderness atmosphere and the pulp and timber interests that wanted a series of dams throughout the Rainy Lake watershed to develop the border country's waterpower potential. The 1941 Arrowhead Association map for tourists suggests the northeastern part of the state's wealth of recreation opportunities, including canoeing and fishing.

The creation of a protected "primitive area" within the Superior National Forest dates to the late 1920s, when Secretary of Agriculture William Jardine issued a proclamation that carved it out of the vast region. Ernest Oberholtzer, whose tireless efforts had pushed the necessary legislation through Congress, was the greatest advocate for the establishment of a pristine wilderness area. Oberholtzer had canoed extensively through the region and respectfully studied the Native American cultures and knowledge of the region. Full protection of the Boundary Waters Canoe Area was not completed until 1964, when the National Wilderness Act was passed and the BWCA became part of the National Wilderness Preservation System.

Although called a wilderness, the BWCAW is not covered by virgin forest. Most of the area was logged at one time and experienced sparse settlement. Trappers, prospectors, and recluses built cabins along the roads through the region. The 1969 national forest map indicates that a few parcels of pri-

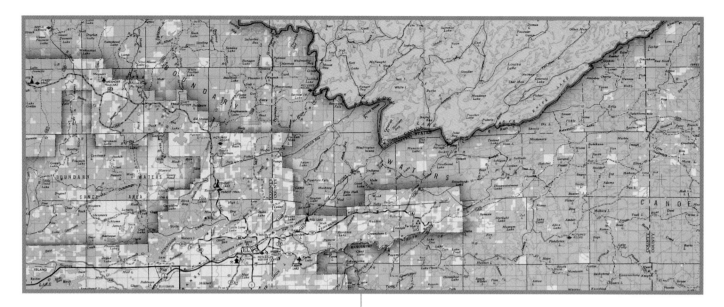

vate land existed inside the boundary of the forest and the BWCAW. It also shows the BWCAW "island" on the southwestern side that is linked to the rest by the Echo Trail. Because the area had been occupied for hundreds of years and had been important to the early economy of the region, there remain a host of conflicts over the use of motors within the boundaries. In 1978, the BWCAW Act of Congress created the present set of regulations, including limiting the use of motorboats to a few large lakes along the boundary. Today the BWCAW is a roadless area, and air travel below 4,000 feet is forbidden. Some people believe that more motorized travel should be allowed so that the landscape can be experienced by more people than just the hardy canoeists and campers who presently enjoy it. Because wintertime use is limited, snowmobilers would like to see trails opened. On the other hand, many canoe campers believe the area is already overused and that the restricted area should be expanded. Over the years, the Boy Scouts and the YWCA and YMCA camps using the area have developed special cultures around the experience of wilderness camping. Some eschew aluminum canoes and use only traditional wooden canoes, because these handcrafted vessels more accurately reflect nature and the nonindustrial world. The watchwords of the BWCAW are simplicity and the importance of preserving the environment.

Lakes Adjacent to the Gunflint Trail, no. 114, *Superior-Quetico Canoe Maps.*

VIRGINIA, MINN.: W. A. FISHER CO., 1952

17 X 22 INCHES

AUTHOR'S COLLECTION

The colorful *Lakes Adjacent to the Gunflint Trail* is one of dozens of detailed maps produced for northern canoeists by the Fisher Map Company of Virginia, Minnesota. Founded in 1922 by William A. Fisher, the company was Minnesota's first color printer north of the Twin Cities. Fisher built his business by developing the most detailed and complete map system of the Minnesota-Canada borderlands ever created. His legacy still exists: the W. A. Fisher Advertising & Printing Map Division continues to provide maps of the lakes and lands of Minnesota to a host of

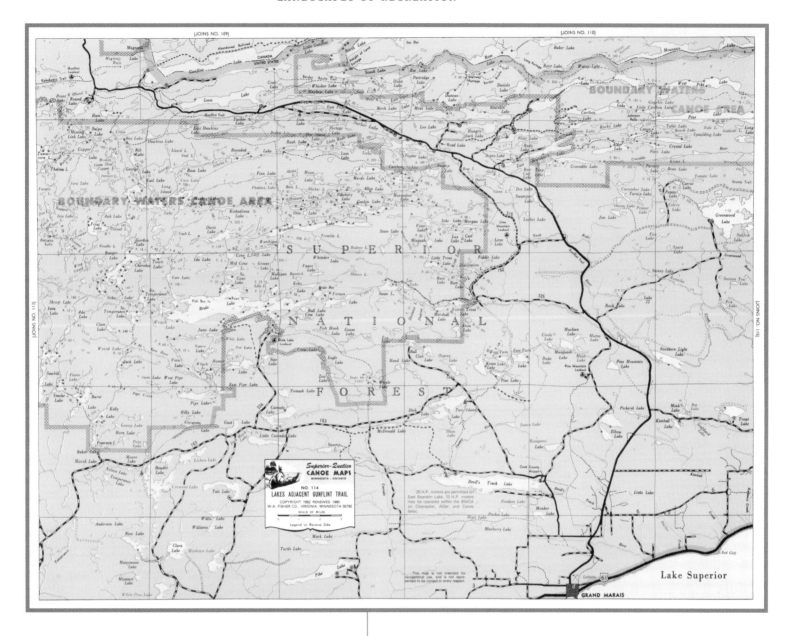

users, ranging from government agencies to BWCAW travelers and other visitors to the North Country. Canoeists treasure these maps, which are frequently annotated with the locations of favorite campsites and memorable events. Well-worn annotated maps keep fresh memories of this fabulous landscape.

The purpose of a canoe map is to provide enough detail so that paddlers can plan their journey through landscapes filled with water and rocks. The design emphasizes the important features—lakes and rivers. Because the maps are intended for use on the water, topographic details are not shown. Instead, the maps feature campsites and the lengths of the portages.

The bright yellow map also illustrates some of the dilemmas facing the BWCAW. The area is a political unit defined with standard property descriptions. Therefore, its boundaries are straight lines. In some places roads come to the immediate edge of the BWCAW without desirable transition or buffer zones. In addition, the fabled Gunflint Trail road penetrates deep into the area, providing, some would say, unduly easy access into the heart of the roadless section.

Ed Langle. *Minneapolis: The City of Lakes.*

VANCOUVER, BRITISH COLUMBIA: TRANSCONTINENTAL CARTOGRAPHERS, 1971

40 X 30 INCHES

AUTHOR'S COLLECTION

This playfully distorted map of Minneapolis from 1971 would have led astray any visitor who attempted to use it, so it is hardly a map in the normal sense. But it presents a spatial image of the city useful to readers curious about the city's natural and man-made features. The map publisher employed the tradition of custom-sales cartography developed a century earlier by the county map and atlas publishers in preparing his map. Just as the publishers of the Andreas *Atlas* added illustrations of properties when farmers paid subscription fees, the establishments shown in this Minneapolis view paid to have their building or corporate logo included in the zany scene, created by artists who blended realistic and fanciful images. The result can be humorous; for example, Hertz Rent A Car dominates the airport. Apparently, airline advertising executives determined that a contribution to this project would not deliver any new customers, and there is no mention of any of the famous airlines associated with the Twin Cities.

The vertical dimensions of the poster map caused some strange scrunching of the land. Southdale mall has been pushed north toward the shore of Lake Harriet; Met Center, home of the North Stars hockey team, has been moved east of Met Stadium. An astonishing shift is the rendering that shows Minnehaha Falls miraculously flowing upstream on its way to the river, which is almost as surprising as the bold assertion that Minneapolis has its own state fair. Auto congestion on the freeways is accurately shown, but freeways are not identified. For some reason, the sign for Nordeast Minneapolis is posted in northwest Minneapolis, and a sign "to the frozen wastes of Canada" is included. Is this an expression of British Columbian attitudes toward the Prairie Provinces? Given the size of the Grain Belt Brewery, which has moved south to dominate the Main Street scene in old St. Anthony, and other mentions of its products around the map, it would seem that this caricature map was designed to be hung behind the wet bars in basement recreation rooms. On the other hand, it might represent a vision of Minneapolis that resulted after imbibing too much of that malt product. Like Grain Belt, many of the other firms shown in this view did not profit enough from their participation and have ceased to exist.

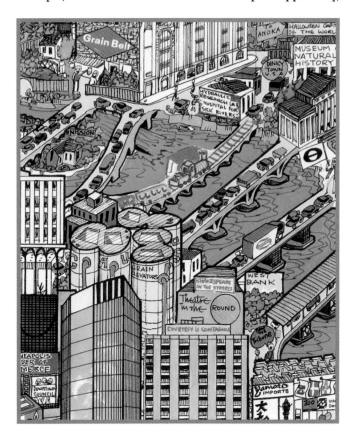

It is hard to know who was expected to buy these maps, which seem to have been marketed as souvenirs for the growing number of tourists beginning to visit the Twin Cities to take advantage of the cultural, sports, shopping, and theater venues available in the core cities. Like travel posters, they were probably designed to promote their places. The colorful and frenetic graphic rendering conveys the impression of a bustling place, although the mapmaker at the same time reserved a great deal of space for rivers, lakes, and parks, the signature landscapes of Minneapolis, "The City of Lakes."

LANDSCAPES OF RECREATION

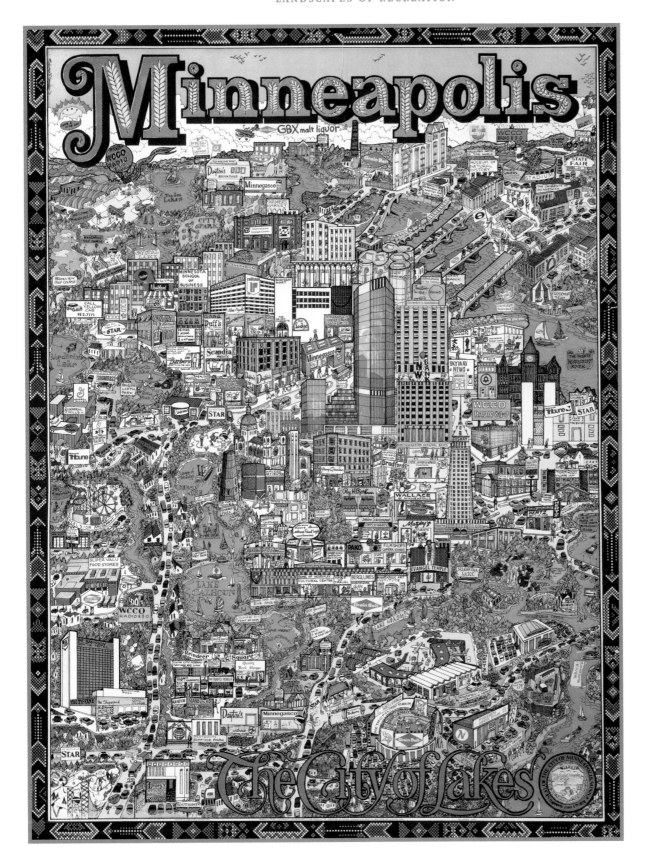

S. Vero. *St. Paul.*

SCARBOROUGH, ONTARIO: ARCHER, 1973

26.75 X 40 INCHES

AUTHOR'S COLLECTION

In contrast to the preceding map of Minneapolis, this illustrated city map of St. Paul from the same era emphasizes history, not businesses and geographic features. The brownish border suggests "old" and "worn," and it is filled with text, in contrast to the bright geometric design bordering the Minneapolis view. The St. Paul mapmaker solved the tricky problem of the city's complicated street system by simply not including most of them. The streets that are represented are nearly devoid of cars and pedestrians. In the city's center, however, a parade progresses down Kellogg Boulevard, preceded by a Winter Carnival Vulcan engaging in the now-forbidden practice of chasing and smudging a young woman.

Buildings selected for inclusion by the mapmaker are nearly equally balanced between commercial structures and cultural landmarks, including colleges, churches, and old housing districts, which would not have paid a fee to be included. In sharp contrast to the Minneapolis view, St. Paul has a winter, though it is only on the east side, where the lakes are frozen hard enough to support car races, fishing contests, and ball games. Like Minneapolis, St. Paul is dominated by a brewery, this one being Hamm's. St. Paul's favorite animal, the Hamm's Bear, is featured in several locations. It is hard to believe that the creator of the view knew St. Paul very well, since he shows bathing beauties and swimmers in the Mississippi at the site of Pig's Eye Waste Treatment Plant.

Of course, it is unnecessary to criticize these two city views on the grounds of spatial accuracy. Although they do reflect the spatial structure of cities, the maps are vehicles for displaying activities, cultural landmarks, and corporate advertising. They were published at a time when the economic viability and future of Minnesota's central cities were in doubt. In some respects, these views were designed to project positive images of cities to the residents themselves, just as local groups promoted a "back-to-the-cities movement" that stressed the historic features of cities, the convenience of living close to work, and cultural amenities.

At the time these views were made, city downtowns had many retail businesses and several corporate headquarters, but planners and entrepreneurs knew the nature of cities was changing and that tourists seeking cultural and historic experiences would become an increasing part of urban economies. Observers also understood that the convention business would become the mainstay of hotels and restaurants, and in order to attract visitors and conventions, cities had to be marketed. In a sense, they had to develop brands and logos, like other consumer products. Some cities created new nicknames, while others, like Minneapolis,

LANDSCAPES OF RECREATION

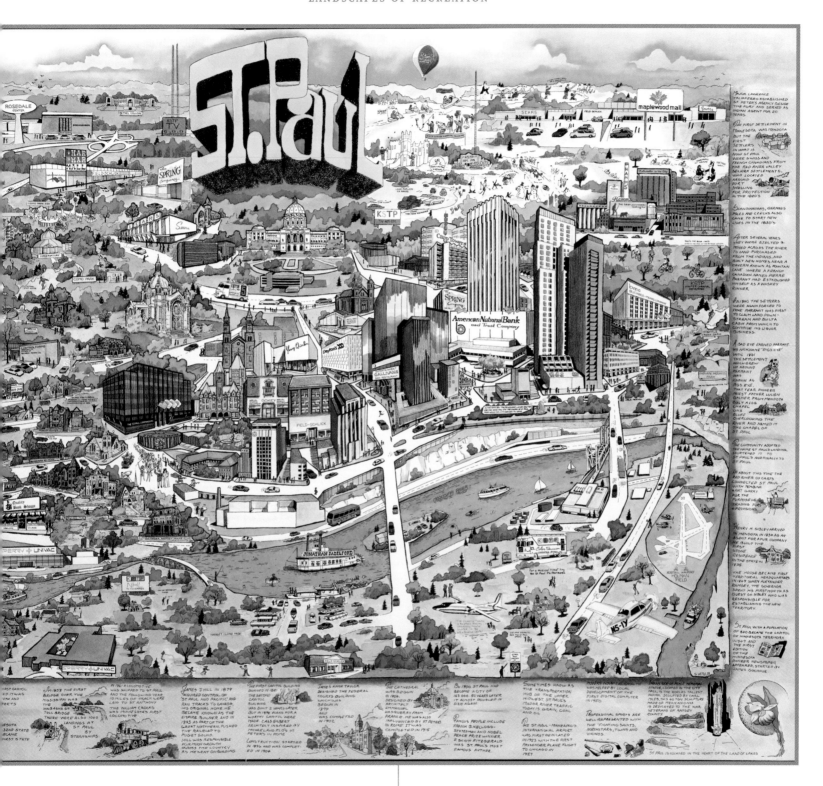

emphasized older names. The smaller city, St. Paul, had not developed a brand equivalent to the "City of Lakes" and struggled to maintain a separate identity from Minneapolis.

Using symbols to convey a sense of a place is as old as cartography itself, and cartoons and caricatures of places may be the most easily remembered.

10 MAPPING THE MODERN LANDSCAPE: TWO MAPS

Cartography has played a critical role in the analysis and use of Minnesota's resources. Over the years, published maps have presented an enormous range of information, from atmospheric conditions to the structure of the earth's crust below the surface. Comparing this very large Minnesota vegetation map and the land-use map that follows suggests this great diversity of content while also demonstrating dramatic changes in the way cartographers make maps.

The great cartographer Francis J. Marschner, who visited Minnesota only in his mind, compiled the best map of the state's preagricultural landscape. The Austrian-born Marschner, who had studied at the Cartographic Institute in Berlin, came to the Bureau of Agricultural Economics in Washington, D.C., where he began reading the 200 mostly handwritten volumes of Minnesota field notes and plat maps made by the public land surveyors between 1847 and 1907. Working from those observations, Marschner combined them with his interpretation of soil and climate data for the state's 2,775 townships and envisioned patterns of vegetation to create one of the most valuable maps of the state. Although it has some limitations, scholars constantly find the map useful and have corroborated its patterns. The map shows the state's three broad diagonal zones of vegetation: the prairie in the southwest, the mixed-hardwoods central zone, and the pine forest of the northeast.

Because the map's vegetation patterns reflect the nature of soils, landforms, climate, and, to some extent, the movements and activities of animals, the map's various vegetation categories can be read as ecological zones. Most accurate in the areas where the patterns are least complex and somewhat less useful in the northern parts of the state, it has served a wide variety of professional and amateur students of the state's ecology and development.

Marschner's map is an excellent example of the cartographer's craft, or art. Using sometimes limited land-survey observations, he generalized to create a whole that is far greater than the sum of the parts. Skill and vision are paramount in this kind of mapmaking.

Unfortunately, little is known today about why the map was made. Marschner worked at a time when conservation and resource management questions were coming to the forefront of many policy debates. Perhaps he thought there would

be a need for such a map. Perhaps he welcomed a serious professional challenge. He must have intended the map for publication, but only two copies were made at the time. The original went to the North Central Forest Experiment Station in

Francis J. Marschner. *The Original Vegetation of Minnesota*, compiled from the U.S. General Land Office Survey notes, U.S. Department of Agriculture, 1930. Redrawn from the original by Patricia Burwell and Sandra J. Hass, Department of Geography, University of Minnesota, under the direction of Miron L. Heinselman.

ST. PAUL:
NORTH CENTRAL FOREST EXPERIMENT STATION, 1974

55 X 49 INCHES

MINNESOTA HISTORICAL SOCIETY

St. Paul, and a copy was kept in the files of Washington. After the Minnesota copy was lost, perhaps to a World War II paper drive, the Washington copy was sent in 1963 to St. Paul, where it sat in a file drawer until geographer Miron Heinselman discovered the map. Working with two cartographers from the University of Minnesota, Heinselman produced the first published edition of the map in 1974.

Marschner himself had retired in 1952. He continued to walk seven miles each day to his office at the Agriculture Department. Already legendary for his accomplishments and diligence, he died in 1966 of a heart attack while walking across the Capitol Mall during a heavy snowstorm at age eighty-three. Without family in the United States, he was buried in an unmarked pauper's grave.

Minnesota Land Use and Cover, 1990s Census of the Land. **Digital map produced by Minnesota Department of Natural Resources.**

ST. PAUL: ASSOCIATION OF MINNESOTA COUNTIES, UNIVERSITY OF MINNESOTA CENTER FOR URBAN AND REGIONAL AFFAIRS, SCIENCE MUSEUM OF MINNESOTA, AND DEPARTMENT OF NATURAL RESOURCES, 1990S

50 X 42 INCHES

AUTHOR'S COLLECTION

This map from the 1990s heralds the new concerns of Minnesota cartography. Maps have long since moved beyond questions of where Minnesota is and now examine in detail the characteristics of regions and places within the state. Early explorers' visionary and artistic maps have been replaced by Geographic Information Science (GIS) maps created from satellite images and digital mapping programs.

This digital map contrasts sharply with the work of Nicollet and the other surveyors who personally traveled around the region to view the landscape and make astronomical observations. It also differs from Marschner's vegetation map and others that were based on the Public Land Survey and added new layers of information based on a cartographer's interpretation of other people's data. It is far different from illustrated tourist maps that are largely meant to create an image or feeling for a place.

This land-use map is derived from 1991–92 satellite imagery. Satellite data is broadcast to the earth's surface, where computers reassemble it in a variety of ways and combine it with other data, such as political boundaries, to create maps. The computer mapping technology used for this map identified eleven different vegetative communities, as small as 5 acres. When applied to the state's fifty-four million acres, this scale makes a phenomenal amount of detail visible.

A comparison of earlier maps with this modern map clearly shows that several great human processes have changed the surface of Minnesota during the past 150 years. The process of agriculture consisted of plowing grasslands, draining marshes and other wetlands, and clearing forests. Lumbering brought the clearing of virgin forest but now involves managing forests for sustained yields. The third process is mining, both on the Iron Range and in sand and gravel mines near urban areas. The fourth major process, urbanization—like agriculture, a process with several components—involves draining land and creating an impervious surface and new vegetation communities. As a result, all but a few fragments of prairie and wetlands in the southwestern portion of the state have been brought under cultivation. And only a few large parcels of forest and wetlands remain in the transition zone of mixed hardwood and prairies. Forest and marshland still cover most of the northern part of the state, but the land has been changed by the lumbering industry, as it is crisscrossed with roads and broken up by cities and some agricultural areas. The huge open-pit mines on the northeastern Iron Range are clearly visible to satellites.

JUST AS MAPS HAVE GUIDED THE DEVELOPMENT of Minnesota over the past five centuries, maps will continue to chart and direct the state's growth. Contemporary GIS mapping systems will enable planning and analysis that is much more inclusive and detailed than we can now imagine, but these maps will, as they have always done, reflect the values of their times to provide future Minnesotans with a glimpse of the state as their predecessors saw it.

MAPPING THE MODERN LANDSCAPE: TWO MAPS

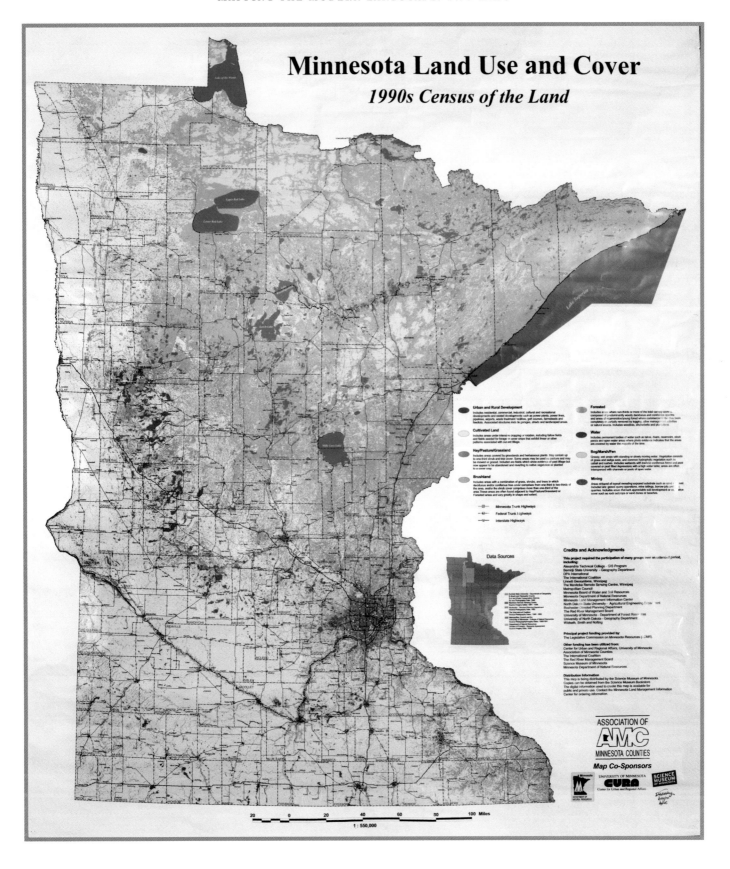

✺ BIBLIOGRAPHY AND SOURCES ✺

(Text notes in parentheses refer to these sources.)

1 ✺ FIRST EUROPEAN VIEWS
(PAGES 6–27)

Armstrong, Joe C. W. *Champlain.* Toronto: Macmillan of Canada, 1987.

Carver, Jonathan. *The Journals of Jonathan Carver and Related Documents, 1766–1770.* Edited by John Parker. St. Paul: Minnesota Historical Society Press, 1976.

Goetzmann, William H., and Glyndwr Williams. *The Atlas of North American Exploration: From the Norse Voyages to the Race to the Pole.* New York: Prentice Hall General Reference, 1992.

Hennepin, Louis. *Father Louis Hennepin's Description of Louisiana.* Translated by Marion E. Cross. Introduction by Grace Lee Nute. Minneapolis: University of Minnesota Press, 1938.

Library of Congress. The Map and Geography Division has posted Waldseemüller's wall map on permanent display at the Web site http://www.loc.gov/rr/geogmap/exh.html

Mealing, S. R., ed. *The Jesuit Relations and Allied Documents: A Selection.* Toronto: Macmillan of Canada, 1969.

Short, John Rennie. *The World through Maps: A History of Cartography.* Buffalo, N.Y.: Firefly Books, 2003.

Upham, Warren. *Minnesota Place Names: A Geographical Encyclopedia.* St. Paul: Minnesota Historical Society, 2001.

Urness, Carol. *Waldseemüller's Globe and Planisphere.* Minneapolis: James Ford Bell Library, 1999.

Waldseemüller, Martin. *Cosmographiae Introductio* (Introduction to Cosmography). Saint Dié, France: Walter and Nikolaus Lud, 1507.

2 ✺ MAPPING AND MEASURING THE LAND
(PAGES 28–37)

The Journals of Joseph N. Nicollet: A Scientist in the Mississippi Headwaters, with Notes on Indian Life, 1836–37. Translated from the French by Andre Ferter. Edited by Martha Coleman Bray. St. Paul: Minnesota Historical Society, 1970.

Schoolcraft, Henry Rowe. *Narrative of an Expedition through the Upper Mississippi to Itasca Lake, the Actual Source of This River: Embracing an Exploratory Trip through the St. Croix and Burntwood (or Broule) Rivers: In 1832.* New York: Harper and Brothers, 1834.

Schubert, Frank. *Vanguard of Expansion: Army Engineers in the Transmississippi West, 1819–1879.* http://www.cr.nps.gov/history/online_books/shubert/foreword.htm

Squires, Rod. "An Introduction to Nicollet's Map of 1843." *Minnesota Surveyor* (Spring 2006): 26–31.

"Zebulon Pike: Hard-Luck Explorer": www.nps.gov/jeff/LewisClark2/Circa1804/WestwardExpansion/EarlyExplorers/ZebulonPike.htm

3 ✺ CLAIMING THE LAND: COMMERCIAL MAP PUBLISHERS
(PAGES 38–61)

Danzer, Gerald A. "George Cram and the American Perception of Space." In *Chicago Map Makers: Essays on the Rise of the City's Map Trade,* edited by Michael P. Conzen, pp. 32–46. Chicago: Chicago Historical Society for the Chicago Map Society, 1984.

Ristow, R. R. "The S. A. Mitchell and J. H. Map Publishing Companies." In R. R. Ristow, *American Maps and Mapmakers: Commercial Cartography in the Nineteenth Century,* pp. 303–26. Detroit: Wayne State University Press, 1985.

White, Bruce. "The Power of Whiteness." *Minnesota History* (Winter 1998–99): 182–83.

4 ✺ OWNING THE LAND: COUNTY ATLASES
(PAGES 62–71)

Conzen, Michael P. "Maps for the Masses: Alfred T. Andreas and the Midwestern County Atlas Map Trade." In *Chicago Mapmakers: Essays on the Rise of the City's Map Trade,* edited by Michael P. Conzen, pp. 47–63. Chicago: Chicago Historical Society, 1984.

Flanagan, Maureen, Cynthia Klinger, Robert Noyed, and Robert Wittman. *Aimless Drift or Quiet Redesign: An Examination of Public Education Governance in Minnesota.* Minneapolis: Bush Educators Program, 2001.

Holmquist, June Drenning, ed. *They Chose Minnesota: A Survey of the State's Ethnic Groups.* St. Paul: Minnesota Historical Society Press, 1981.

Kiefer, Monica. *American Children through Their Books, 1700–1835.* Philadelphia: University of Pennsylvania Press, 1948.

Ristow, R. R. "The County Atlas." In R. R. Ristow, *American Maps and Mapmakers: Commercial Cartography in the Nineteenth Century,* pp. 403–26. Detroit: Wayne State University Press, 1985.

Thrower, Norman. "The County Atlas of the United States." *Surveying and Mapping; Quarterly Publication of American Congress on Survey and Mapping* 23 (1963): 365–73.

Thrower, Norman J. W. "Cadastral Survey and County Atlases of the United States." *Cartographic Journal* 9, no. 1 (June 1972): 43–51.

Treude, Mai. *Windows to the Past: A Bibliography of Minnesota County Atlases.* Edited by Pamela Espeland and Judith H. Weir. Minneapolis: University of Minnesota, Center for Urban and Regional Affairs, 1980.

Upham, Warren. *Minnesota Place Names: A Geographical Encyclopedia.* St. Paul: Minnesota Historical Society Press, 2001.

5 ✻ MAPPING THE STATE: THE ANDREAS ILLUSTRATED ATLAS (PAGES 72–81)

Conzen, Michael P. "Maps for the Masses: Alfred T. Andreas and the Midwestern County Atlas Trade." In *Chicago Mapmakers: Essays on the Rise of the City's Map Trade,* edited by Michael P. Conzen, pp. 47–63. Chicago: Chicago Historical Society, 1984.

Miller, Dana. "The Vermilion Lake Gold Rush, 1865–1866." In *Entrepreneurs and Immigrants,* edited by Michael G. Karri, n. p. Chisholm: Iron Range Research Center, a Division of the Iron Range Resources and Rehabilitation Board, 1991.

Ristow, R. R. "The New State of State Atlases." In R. R. Ristow, *American Maps and Mapmakers: Commercial Cartography in the Nineteenth Century,* pp. 427–444. Detroit: Wayne State University Press, 1985.

6 ✻ CITY PLATS AND MAPS (PAGES 82–109)

Alanen, Arnold R. "Morgan Park: U.S. Steel and a Planned Company Town." In *Duluth Sketches of the Past, a Bicentennial Collection,* edited by Ryck Lydecker and Lawrence J. Sommer, pp. 111–25. Duluth, Minn.: American Revolution Bicentennial Commission, 1976.

Borchert, John R. *America's Northern Heartland: An Economic and Historical Geography of the Upper Midwest.* Minneapolis: University of Minnesota Press, 1987.

Conzen, Michael P. "The Non-Pennsylvania Town: Diffusion of Urban Plan Forms in the American West." *Geographical Review* 96 (2006): 183–211.

Gimmestad, Dennis. "Territorial Space: Platting New Ulm." *Minnesota History* 56, no. 6 (Summer 1999): 345–50.

History of Winona, Dodge, and Olmsted Counties, pp. 934–35. Chicago: H. H. Hill and Company, 1884.

Johnson, Hildegard Binder. "The Founding of New Ulm, Minnesota." *American German Review* (June 1946): 8–12.

Reps, John W. "Bird's-Eye Views of the Upper River by Albert Ruger and His Followers, 1867–1879." In John R. Reps, editor, *Cities of the Mississippi: Nineteenth-Century Images of Urban Development.* Columbia: University of Missouri Press, 1994.

Schmieder, T. "Civic Geometry: Frontier Forms of Minnesota's County Seats." *Minnesota History* 57 (2001): 330–45.

Tyler, Alice F. "William F. Pfaender and the Founding of New Ulm." *Minnesota History* 30 (1949): 24–35.

Vollmar, Reiner, Free University of Berlin. Personal Correspondence, January 2007.

7 ✻ MAPPING THE TRANSPORTATION CONNECTIONS (PAGES 110–129)

Bond, J. W. *Minnesota and Its Resources.* New York: Redfield, 1853.

Jackson, W. Turrentine. "Minnesota Territory, an Experimental Laboratory, 1850–1." In *Wagon Roads West: A Study of Federal Road Surveys and Construction in the Transmississippi West, 1846–1869.* New Haven: Yale University Press, 1952.

Larson, Arthur J. *The Development of the Minnesota Road System.* St. Paul: Minnesota Historical Society, 1966.

Peters, Cynthia. "Rand, McNally in the Nineteenth Century, Reaching for a National Market." In *Chicago Mapmakers: Essays on the Rise of the City's Map Trade,* edited by Michael P. Conzen, pp. 64–72. Chicago: Chicago Historical Society, 1984.

Ristow, R. R. "Rand McNally & Co." In R. R. Ristow, *American Maps and Mapmakers: Commercial Cartography in the Nineteenth Century,* pp. 467–81. Detroit: Wayne State University Press, 1985.

Seavey, Charles A. "The Wagon Road Surveys." In *Exploration and Mapping of the American West.* Map and Geography Round Table of the American Library Association, Occasional Paper #1, edited by Donna P. Koepp. Chicago: Speculum Orbis Press, 1986.

Stevens, Isaac I. Report of Exploration of a Route for the Pacific Railroad, Near the Forty-seventh and Forty-ninth Parallels, from St. Paul to Puget Sound, 33d Cong., 1st sess., House of Representatives Executive Document 129, I, pp. 4–5.

8 ✻ MAPPING THE DEVELOPING TWIN CITIES (PAGES 130–175)

Bennett, Edward, and Andrew Crawford. *Plan of Minneapolis.* Minneapolis: Civic Commission, 1917.

Folwell, William Watts. *A History of Minnesota.* Vol. 1. St. Paul: Minnesota Historical Society, 1956.

Holland, Robert A. *Chicago in Maps: 1612 to 2000.* New York: Rizzoli, 2005.

Lanegran, David A. "Swedes in the Twin Cities." In *Swedes in Minnesota,* edited by Philip J. Anderson and Dag Blanck, pp. 39–56. St. Paul: Minnesota Historical Society Press, 2001.

———. *St. Anthony Park: Portrait of a Community.* St. Paul: District 12 Community Council and St. Anthony Park Association, 1987.

Lanegran, David A., John Borchert, David Gebhart, and Judith Martin. *The Legacy of Minneapolis: Preservation amid Change.* Minneapolis: Voyager Press, 1983.

Lanegran, David A., and Judith Martin. *Where We Live: Residential Districts of the Twin Cities.* Minneapolis: University of Minnesota Press, 1983.

Lanegran, David A., and Ernest Sandeen. *The Lake District of Minneapolis: A Neighborhood History.* St. Paul: Living Historical Museum, 1979.

Minneapolis Planning and Economic Development Department. *Metro '85: Study of Development of Program and Priorities for Expanded Jobs and Investment Opportunities in Central Minneapolis.* Minneapolis: Minneapolis Planning and Economic Development Department, 1970.

Reps, John W. *Cities of the Mississippi: Nineteenth-Century Images of Urban Development.* Columbia: University of Missouri Press, 1994.

Schmid, Calvin F. *Social Saga of Two Cities: An Ecological and Statistical Study of Social Trends in Minneapolis and St. Paul.* Minneapolis: Minneapolis Council of Social Agencies, 1937.

9 ❋ LANDSCAPES OF RECREATION (PAGES 176–197)

Lanegran, David A. "Minnesota: Nature's Playground." In *Minnesota: A Different America. Daedalus, Journal of the American Academy of Arts and Sciences* 129 (Summer 2000): 81–100.

10 ❋ MAPPING THE MODERN LANDSCAPE: TWO MAPS (PAGES 198–201)

Brady, Tim. "The Mystery of a Map and Man." http://www.dnr.state.mn.us/volunteer/janfeb03/mystery.html

COMPLETE LIST OF MAPS

(with repositories)

1 ✸ FIRST EUROPEAN VIEWS
(PAGES 6–27)

Martin Waldseemüller. The globe gores. St. Dié [France], 1507. 7.5 x 14 inches. James Ford Bell Library, University of Minnesota.

Samuel de Champlain. *Carte de la Nouvelle France . . .* , from *Les Voyages de la Nouvelle France occidentale, dicte Canada* [Map of New France, from The Voyages of Western New France, Called Canada]. Paris: Le-Mur, 1632. 20.75 x 17 inches. James Ford Bell Library, University of Minnesota.

Claude Dablon and Claude Jean Allouez. *Lac Superieur et autres lieux ou sont les missions des peres de la Compagnie de Jesus . . .* , from *Relation de ce qui s'est passé de plus remarquable aux missions des peres de la Compagnie de Jesus en la Nouvelle France, les années 1670 & 1671* [Lake Superior and Other Places Where the Missions of the Fathers of the Society of Jesus Are Located, from Account of the Most Remarkable Events That Happened at the Missions of the Fathers of the Society of Jesus in New France in the Years 1670 and 1671]. Paris: Sebastien Mabre-Cramoisy, 1672. 14 x 18.5 inches. James Ford Bell Library, University of Minnesota.

Louis Hennepin. *Carte de la Nouvelle France et de la Louisiane nouvellement decouverte au Sud Quest de la Nouvelle France,* from *Description de la Louisiane* [Map of New France and the Newly Discovered Louisiana in the Southwest of New France, from A Description of Louisiana]. Paris: Veuve Sebastien Huré, 1683. 10.2 x 17.3 inches. Minnesota Historical Society.

Vincenzo Coronelli. *Partie occidentale du Canada ou de la Nouvelle France* [The Western Part of Canada or of New France]. Paris: J. B. Nolin, 1688. 16.9 x 22.8 inches. Minnesota Historical Society.

Henri Abraham Chatelain. *Carte particuliere du fleuve Saint Louis,* from *Atlas historique* [A Distinctive Map of the St. Louis River, from Historical Atlas], vol. 6. Amsterdam: Chatelain, 1719[?]. 16 x 19.5 inches. Minnesota Historical Society.

Johann Baptist Homann. *Amplissimae regionis Mississipi seu provinciae Ludovicianai* [Large Map of the Mississippi Region or Louisiana Province]. Nuremberg, Germany: Homann, c. 1720. 20.5 x 24 inches. Minnesota Historical Society.

A New Map of North America from the Latest Discoveries, from *London Magazine* 32 (February): 64. London, 1763. 10.6 x 15 inches. Minnesota Historical Society.

John Mitchell. *A Map of the British and French Dominions in North America.* London: J. Mitchell for Jeffrys and Faden [1774]. 8 sheets. 6.5 x 4.5 feet (assembled). James Ford Bell Library, University of Minnesota.

A Plan of Captain Carver's Travels in the Interior Parts of North America in 1766 and 1767, from Jonathan Carver, *Travels through the Interior Parts of North-America, in the Years 1766, 1767 and 1768.* London: For the Author, 1779. 12.5 x 17 inches. Minnesota Historical Society.

2 ✸ MAPPING AND MEASURING THE LAND
(PAGES 28–37)

Map of the Mississippi River from the Source to the Mouth of the Missouri, from Zebulon Montgomery Pike, *An Account of Expeditions to the Source of the Mississippi River.* Philadelphia: C. & A. Conrad & Co., 1810. 12.4 x 30 inches. Minnesota Historical Society.

James Allen. *Map of the Route Passed Over by an Expedition into the Indian Country in 1832 to the Source of the Mississippi.* Made by Lieutenant Drayton from the original draft drawn by topographical engineer Lieutenant James Allen, from *American State Papers, Documents, Legislative and Executive, of the Congress of the United States, . . . Commencing March 15, 1832, and Ending January 5, 1836.* Vol. 5. Washington, D.C.: Military Affairs (Serial set 020, p. 312), 1860. 15 x 19 inches. Minnesota Historical Society.

Joseph Nicolas Nicollet. *Hydrographical Basin of the Upper Mississippi River: From Astronomical and Barometrical Observations, Surveys, and Information, by J. N. Nicollet, in the Years 1836, 1837, 1838, 1839, and 1840; Assisted in 1838, 1839 & 1840 by Lieut. J. C. Fremont, of the Corps of Topographical Engineers, 1843.* Washington, D.C., 1843. 36.8 x 30.4 inches. Minnesota Historical Society.

Survey of Township 103 North, Range 19 West, Freeborn County. Washington, D.C., 1855. 10 x 14 inches. Minnesota Historical Society.

3 ✸ CLAIMING THE LAND: COMMERCIAL MAP PUBLISHERS
(PAGES 38–61)

John Farmer. *Improved Map of the Territories of Michigan and Ouisconsin on a Scale of 30 Geographical Miles to an Inch.* New York: J. H. Colton & Co., 1835. 24.5 x 32 inches. Minnesota Historical Society.

J. Ayers, compiler and engraver. *Map of the Settled Part of Wisconsin Territory, Compiled from the Latest Authorities,* from William Rudolph Smith, *Observations on the Wisconsin Territory: Chiefly on That Part Called the Wisconsin Land District.* Philadelphia: E. L. Carey & A. Hart, 1838. 22 x 17.75 inches. Minnesota Historical Society.

Sidney E. Morse and Samuel Breese. *Iowa and Wisconsin Chiefly from the Map of J. N. Nicollet, 1844,* from *Morse's North American Atlas.* New York: Harper Bros., 1845. 14.5 x 18 inches. Minnesota Historical Society.

H. N. Burroughs. *United States,* from Augustus Mitchell, *New Universal Atlas.* Philadelphia: 1848 or 1849. 17 x 14 inches. Author's Collection.

Map of the Organized Counties of Minnesota. With an inset *Map of Minnesota Territory.* Philadelphia: Thomas Cowperthwait & Co., 1850. 22 x 17.75 inches. Minnesota Historical Society.

J. H. Young. *Map of Minnesota Territory* (copyright 1850), no. 36, from *Mitchell's Universal Atlas.* Philadelphia: Thomas Cowperthwait & Co., 1852. 13.75 x 17.6 inches. Author's Collection.

J. H. Young. *Map of Minnesota Territory* (copyright 1850), no. 36, from *Mitchell's Universal Atlas[?].* Philadelphia: Cowperthwait, De Silver and Butler, 1854. 14.5 x 17.5 inches. Author's Collection.

Minnesota, no. 49, from *Colton's Atlas of the World.* New York: J. H. Colton, 1855. 14.5 x 17.5 inches. Author's Collection.

Minnesota, from *The World in Miniature, America.* New Orleans: Morse & Gaston and A. B. Griswold, 1857. 7.5 x 6.25 inches. Author's Collection.

Colton's Minnesota and Dakota (copyright 1855), from *Colton's Atlas of the World.* New York: J. H. Colton, 1858[?]. 12.25 x 15.25 inches. Author's Collection.

Johnson's Minnesota, from *Johnson's New Illustrated (Steel Plate) Family Atlas.* New York: A. L. Johnson, 1865. 18 x 14 inches. Author's Collection.

County Map of Minnesota, drawn and engraved by William H. Gamble, no. 81, from *Mitchell's New General Atlas.* Philadelphia: S. Augustus Mitchell, 1877. 15.25 x 12.25 inches. Author's Collection.

4 ✸ OWNING THE LAND: COUNTY ATLASES
(PAGES 62–71)

Plat Book of Chisago County, Minnesota. Minneapolis: C. M. Foote and Co., 1888. 17.5 x 14.25 inches. Minnesota Historical Society.

Whitewater Township, from *Map of Winona County.* Chicago: Lyman G. Bennett and A. C. Smith, 1867[?]. 11.5 x 11 inches. Minnesota Historical Society.

Shafer Township, from *Plat Book of Chisago County, Minnesota.* Minneapolis: C. M. Foote and Co., 1888. 17.5 x 14.25 inches. Minnesota Historical Society.

Swede Prairie Township, from *Farmers' Atlas & Directory of Yellow Medicine County, Minnesota.* St. Paul: *Farmer Magazine* (Webb Publishing), 1913. 16 x 11.5 inches. Minnesota Historical Society.

Cottage Grove Township, from *Plat Book of Washington County.* Minneapolis: Northwest Publishing, 1901. 18.5 x 15 inches. Minnesota Historical Society.

5 ✸ MAPPING THE STATE: THE ANDREAS ILLUSTRATED ATLAS
(PAGES 72–81)

Anoka County, from *Illustrated Historical Atlas of Minnesota.* Chicago: A. T. Andreas, 1874. 17.5 x 14 inches. Minnesota Historical Society.

Counties of Pine, Kanabec, Isanti, and Chisago, from *Illustrated Historical Atlas of Minnesota.* Chicago: A. T. Andreas, 1874. 17.5 x 14 inches. Minnesota Historical Society.

Dakota County, from *Illustrated Historical Atlas of Minnesota.* Chicago: A. T. Andreas, 1874. 17.5 x 14 inches. Minnesota Historical Society.

Northern Minnesota, from *Illustrated Historical Atlas of Minnesota.* Chicago: A. T. Andreas, 1874. 17.5 x 38 inches. Minnesota Historical Society.

6 ✸ CITY PLATS AND MAPS
(PAGES 82–109)

Red Wing, Goodhue County, from *Illustrated Historical Atlas of Minnesota.* Chicago: A. T. Andreas, 1874. 14 x 17.5 inches. Minnesota Historical Society.

Preston, Fillmore County, from *Illustrated Historical Atlas of Minnesota.* Chicago: A. T. Andreas, 1874. 7 x 14 inches. Minnesota Historical Society.

Rochester, Olmsted County, from *Illustrated Historical Atlas of Minnesota.* Chicago: A. T. Andreas, 1874. 17.5 x 14 inches. Minnesota Historical Society.

Heinze Bros. *Map of the City of Brainerd, Crow Wing County, State of Minnesota.* Minneapolis: Johnson, Smith & Harrison, 1883. 42 x 30 inches. Minnesota Historical Society.

Litchfield, Meeker County, from *Illustrated Historical Atlas of Minnesota.* Chicago: A. T. Andreas, 1874. 17.5 x 8 inches. Minnesota Historical Society.

Hibbing Quadrangle, Minnesota. 7.5 Minute Series Topographic, U.S. Department of Interior Geological Survey, 1957. 27 x 22 inches. Author's Collection.

New Ulm, Brown County, Minnesota (plat map), manuscript copy of Christian Prignitz's 1858 plat map, certified accurate by Fred Pfaender, son of Wilhelm Pfaender and Brown County register of deeds in 1896. 27 x 44 inches. Brown County Recorder's Office.

H. G. Schapekahm. *Map of New Ulm, Brown County, Minnesota.* New Ulm: H. G. Schapekahm, 1875. 24 x 38 inches. Author's Collection.

H. Wellge. *Perspective Map of Duluth, Minnesota.* Duluth: Duluth News Co., 1887. 18.5 x 41.5 inches. Minnesota Historical Society.

Frederick B. Roe. *Albertson's Map of the City of Duluth, St. Louis County, Minnesota, and Vicinity.* Duluth: Albertson & Chamberlain, c. 1891. 22 x 43.5 inches. Minnesota Historical Society.

Map of Duluth (Minnesota) and Superior (Wisconsin), Showing Lines of the Duluth Street Railway Company. Duluth: Duluth Street Railway Co., 1917. 17 x 15 inches. Northeast Minnesota Historical Center, University of Minnesota, Duluth.

Morgan Park, Duluth, Minnesota. Dean and Dean Architects, Morrell and Nichols Landscape Architects, Owen Brainerd Consulting. From *Morgan Park Bulletin,* 1917. 7.5 x 24 inches. Northeast Minnesota Historical Center, University of Minnesota, Duluth.

Beaver, from *Plat Book of Winona County, Minnesota.* Minneapolis: C. M. Foot and J. W. Henion, 1894. 3 x 5 inches. Minnesota Historical Society.

Map of Wasioja Township, from *The County of Dodge, Minnesota.* St. Paul: R. L. Polk & Co., 1905. 16 x 12.5 inches. Minnesota Historical Society.

7 MAPPING THE TRANSPORTATION CONNECTIONS (PAGES 110–129)

Map of the General Government Roads in the Territory of Minnesota, September 1854, accompanying the report of the Bureau of Topographical Engineers, Senate Executive Documents, no. 1, 2nd sess., 33rd Congress. 19.5 x 13.25 inches. Author's Collection.

Map of Minnesota Territory, from John Wesley Bond, *Minnesota and Its Resources.* New York: Redfield, 1853. 12.25 x 15 inches. Minnesota Historical Society.

Isaac Ingalls Stevens. *Preliminary Sketch of the Northern Pacific Rail Road: Exploration and Survey, Map 1, "From St. Paul to Riviere des Lacs."* Philadelphia: Wagner and McGuigan Lith., 1853–1854. 26 x 37 inches. Minnesota Historical Society.

Minnesota, from *An Atlas of the Northwest.* Chicago: Rand McNally, 1896. 18.9 x 25.75 inches. Minnesota Historical Society.

Auto Trails Map of Minnesota and Western Wisconsin, from *Commercial Atlas of America.* Chicago: Rand McNally, 1920. 28.25 x 20.75 inches. Minnesota Historical Society.

Early privately published and official state road maps of Minnesota. Craig Solomonson Collection.

State Highway Department Map of Minnesota Showing Status of Improvement of State Roads, January 1919. St. Paul: Minnesota Highway Department, 1919. 19 x 15 inches. Minnesota Historical Society.

8 MAPPING THE DEVELOPING TWIN CITIES (PAGES 130–175)

Jules Guerin. *The Coming Metropolis: General Perspective Looking Northwest Showing the Development of Minneapolis and Connections with Saint Paul and Surrounding Country,* from David Bennett and Andrew Crawford, *Plan of Minneapolis.* Minneapolis: Civic Commission, 1917. 12.25 x 18.5 inches. Author's Collection.

Lieutenant J. L. Thompson. *Map of the Military Reserve, Embracing Fort Snelling.* 1839; copied by William Gordon in 1853. 21 x 21 inches. Hennepin County Surveyor's Office.

Geo. H. Ellsbury. *St. Paul, Minnesota.* Lithographed by Hoffman. Chicago: Chas. Schober & Co., 1874. 16.5 x 29 inches. Minnesota Historical Society.

Plan of St. Anthony Park, a Suburban Addition to St. Paul and Minneapolis. Chicago: Cleveland and French, 1873. 10 x 15 inches. Ramsey County Historical Society.

Report of the Leading Business Houses of Minneapolis, Minnesota. Minneapolis: Augustus Hageboeck, 1886. 10.8 x 16.4 inches. Minnesota Historical Society.

Lowry Hill, from *Insurance Maps of Minneapolis,* vol. 3, pp. 217–18. New York: Sanborn Map Co., 1912. 27.5 x 39.5 inches. Minnesota Historical Society.

Downtown Minneapolis, from *Insurance Maps of Minneapolis,* vol. 3, pp. 257–58. New York: Sanborn Map Co., 1912. 29.5 x 17 inches. Minnesota Historical Society.

Section 32, Township 29, Range 22, plate 14, from *Atlas of the City of St. Paul.* Chicago: Ruben H. Donnelley, 1892. 29.5 x 34 inches. Author's Collection.

St. Paul, Township 28 North, Range 22 West, and *Sub Plan #3,* plates 52 and 53, from *Hopkins Plat Book of St. Paul and Suburbs.* Philadelphia: G. M. Hopkins Co., 1926. 29.5 x 17 inches. Dakota County Historical Society.

The Twin Cities: Their Famous Lakes, River, Parks and Resorts, from *Annual Report of the Twin City Rapid Transit Company.* Minneapolis: Twin City Rapid Transit Company, 1909. (Reprinted to commemorate the dedication of the John R. Borchert Map Library, University of Minnesota, May 26, 1989. Minneapolis: University of Minnesota, 1989.) 10 x 33 inches. Minnesota Historical Society.

Minneapolis West. U.S. Geological Survey edition of 1901 (reprinted 1915). 20 x 16.25 inches. Author's Collection.

Map of Minneapolis, Minnesota, Showing Park System as Recommended by Prof. H. W. S. Cleveland, 1883, from H. W. S. Cleveland, *Suggestions for a Systems of Parks and Parkways for the City of Minneapolis, Read at Meeting of the Minneapolis Park Commissioners, June 2, 1883.* Minneapolis: Johnson, Smith and Harrison, 1883. 8.7 x 10 inches. Minneapolis Public Library.

Map of Proposed Minnehaha Park and Parkways, from *The Aesthetic Development of the United Cities of St. Paul and Minneapolis.* An address given to the Minneapolis Society of Fine Arts, April 2, 1888. 4 x 5 inches. Minneapolis Public Library.

Jules Guerin. *The Sixth Avenue Artery,* from Edward H. Bennett and Andrew Crawford, *Plan of Minneapolis.* Minneapolis: Civic Commission, 1917. 8.25 x 4.6 inches. Author's Collection.

Jules Guerin. *The Sixth Avenue Approach to the Institute of Arts, through Washburn Park,* from Edward H. Bennett and Andrew Crawford, *Plan of Minneapolis.* Minneapolis: Civic Commission, 1917. 6.5 x 10 inches. Author's Collection.

St. Paul Zoning Map, 1922, from City Planning Board, Edward H. Bennett and William E. Parsons, Consultant City Planners, and George H. Herrold, City Plan Engineer, *Plan of St. Paul: The Capital City of Minnesota.* St. Paul: Commissioner of Public Works, 1922. 17 x 23.5 inches. Author's Collection.

Preliminary City Plan, from City Planning Board, Edward H. Bennett and William E. Parsons, Consultant City Planners, and George H. Herrold, City Plan Engineer, *Plan of St. Paul: The Capital City of Minnesota.* St. Paul: Commissioner of Public Works, 1922. 17 x 23.5 inches. Author's Collection.

Natural Areas, Central Segment, St. Paul, 1935, from Calvin Schmid, *Social Saga of Two Cities: An Ecological and Statistical Study of Social Trends in Minneapolis and St. Paul,* p. 181. Minneapolis: Minneapolis Council of Social Agencies, 1937. 8.5 x 11 inches. Author's Collection.

Natural Areas, Middle Segment, Minneapolis, 1935, from Calvin Schmid, *Social Saga of Two Cities: An Ecological and Statistical Study of Social Trends in Minneapolis and St. Paul.* Minneapolis: Minneapolis Council of Social Agencies, 1937, p. 38. 8.5 x 11 inches. Author's Collection.

Vice Areas, Minneapolis, 1936, from Calvin Schmid, *Social Saga of Two Cities: An Ecological and Statistical Study of Social Trends in Minneapolis and St. Paul.* Minneapolis: Minneapolis Council of Social Agencies, 1937, p. 363. 8.5 x 11 inches. Author's Collection.

Souvenir Guide Map of Saint Paul, Showing Places of Historic, Scenic and General Interest. 1953[?]. 8.5 x 17 inches. Ramsey County Historical Society.

Predominant Land Use 1958. St. Paul: Twin Cities Metropolitan Planning Commission, 1959[?]. 22.3 x 17.5 inches. Author's Collection.

Twin Cities Metropolitan Area Generalized Land Use 1968. St. Paul: Metropolitan Council of the Twin Cities Area, January 1968. 22 x 28.75 inches. Author's Collection.

Constellation Cities, 1985, from *Report Number 5,* April 1968. St. Paul: Joint Program, an Interagency Land Use–Transportation Planning Program for the Twin Cities Metropolitan Area, 1968. 22 x 17 inches. Author's Collection.

Twin Cities, 2000 AD: Alternative Patterns, from *Report Number 4,* January 1967. St. Paul: Joint Program, an Interagency Land Use–Transportation Planning Program for the Twin Cities Metropolitan Area, 1967. 11 x 8.5 inches (each). Author's Collection.

Metro Center '85 Illustrative Site Plan, from *Metro Center '85 Study for the Development of Program and Priorities for Expanded Job and Investment Opportunities in Central Minneapolis,* p. 152. Minneapolis: Planning and Development, 1970. 11 x 11 inches. Author's Collection.

Riverfront Housing and Cultural Center Illustrative Site Plan, from *Metro Center '85 Study for the Development of Program and Priorities for Expanded Job and Investment Opportunities in Central Minneapolis,* pp. 129–30. Minneapolis Planning and Development, 1970. 5.5 x 11 inches. Author's Collection.

9 ❋ LANDSCAPES OF RECREATION
(PAGES 176–197)

Langwith's Pictorial Minnesota: The Land of 10,000 Lakes. St. Paul: McGill-Warner, 1921. Author's Collection.

"Minnesota, Where Dreams Come True," cover of *Minnesota Invites You to Live, Work, Play, in the Playground of 10,000 Lakes.* St. Paul: Minnesota Department of Conservation and Minnesota Tourist Bureau, c. 1933. Laura Kigin Collection.

Minnesota Invites You to Live, Work, Play, in the Playground of 10,000 Lakes. St. Paul: Minnesota Department of Conservation and Minnesota Tourist Bureau, c. 1933. 31 x 21 inches. Laura Kigin Collection.

Langwith's Pictorial Map of Minnesota—Come-to-Minnesota Club. Minneapolis: Langwith Map Co., 1920s. 37.75 x 24 inches. Author's Collection.

Alexandria Lake Region in Minnesota. Virginia, Minn.: Fisher Map Co. and Alexandria Chamber of Commerce, 1939. 17.6 x 22.4 inches. Minnesota Historical Society.

Itasca County, Minnesota: The Heart of the 10,000 Lakes Country. St. Paul: McGill-Warner Co., 1923. 18 x 16 inches. Laura Kigin Collection.

Selected Minnesota postcards. Laura Kigin Collection, Author's Collection and Lured to the Lake.

Minnesota State Parks, State Forests, and Recreational Areas. W. H. Wettschreck. St. Paul: Minnesota Department of Conservation, 1941. 30 x 21 inches. Minnesota Historical Society.

Map of the Minnesota Arrowhead Country. St. Paul: Minnesota Arrowhead Association, 1941. 16 x 26.5 inches. Author's Collection.

Superior National Forest. Duluth: U.S. Department of Agriculture, Forest Service, 1969. 26.5 x 48 inches. Author's Collection.

Lakes Adjacent to the Gunflint Trail, no. 114, *Superior-Quetico Canoe Maps.* Virginia, Minn.: W. A. Fisher Co., 1952 and 1980. 17 x 22 inches. Author's Collection, with permission of the publisher.

Ed Langle. *Minneapolis: The City of Lakes.* Vancouver, British Columbia: Transcontinental Cartographers, 1971. 40 x 30 inches. Author's Collection.

S. Vero. *St. Paul.* Scarborough, Ontario: Archer, 1973. 26.75 x 40 inches. Author's Collection.

10 ✹ MAPPING THE MODERN LANDSCAPE: TWO MAPS (PAGES 198–201)

Francis J. Marschner. *The Original Vegetation of Minnesota,* compiled from the U.S. General Land Office Survey notes, Office of Agricultural Economics, U.S. Department of Agriculture, 1930. Redrawn from the original by Patricia Burwell and Sandra J. Hass, cartographers, Department of Geography, University of Minnesota, under the direction of Miron L. Heinselman. St. Paul: North Central Forest Experiment Station, 1974. 55 x 49 inches. Minnesota Historical Society.

Minnesota Land Use and Cover, 1990s Census of the Land. Digital map produced by Minnesota Department of Natural Resources. St. Paul: Association of Minnesota Counties, University of Minnesota Center for Urban and Regional Affairs, Science Museum of Minnesota, and Department of Natural Resources, 1990s. 50 x 42 inches. Author's Collection.

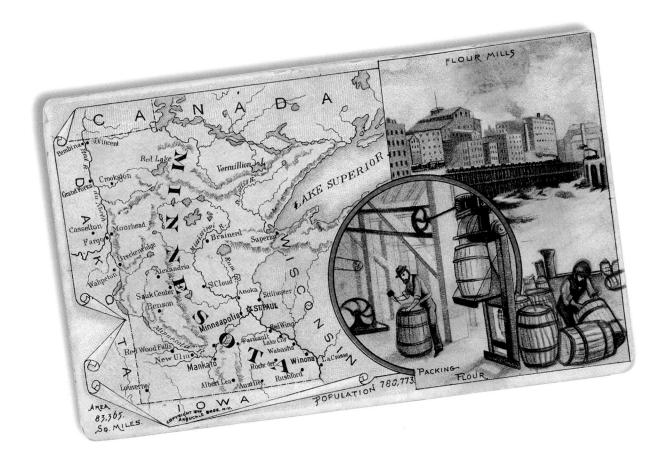

✸ INDEX OF MAPS, PEOPLE, AND PLACES ✸

(titles of maps appear in italics)

A

Aesthetic Development of the United Cities of St. Paul and Minneapolis, 156
Albert Lea, 86
Albertson's Map of the City of Duluth, St. Louis County, 101–2
Alexandria Lake Region in Minnesota, 182–83
Allen, James, *Map of the Route Passed over by an Expedition into Indian Country,* 32, 34
Allouez, Claude, *Lac Superieur,* 12–13
Andreas, Alfred Theodore, 63, 72–81; Minnesota atlas, contents and production of, 73–74
Annual Report of the Twin City Rapid Transit Company, 148
Anoka County, 56, 76
Anoka County, from *Illustrated Historical Atlas of Minnesota,* 74–75
Antoncich, Frank, 182
Arrowhead Region, 55, 180–81, 187, 190–91
Austin, 178
Auto Trails Map of Minnesota and Western Wisconsin, 120–21
Ayers, J., *Map of the Settled Part of Wisconsin Territory,* 42–43

B

Babcock, Charles M., 126
Barn Bluff, 84
Beard, Reverend Henry, 141
Beaver, 64–65; *Beaver, Map of,* 106–7
Beinhorn, Ferdinand, 97
Bennett, Edward H., *Plan of Minneapolis,* 131, 157–59; *Preliminary City Plan,* 162–63; *St. Paul Zoning Map,* 160–61
Bennett, Lyman G., 64
Big Legs, 48, 50
Big Sandy Lake, 32
Big Sioux River, 112
Big Stone Lake, 48, 50
Black Dog, 48, 50
Blue Earth, 49, 53
Blue Earth River, 41

Bond, John Wesley, 114–15
Bottineau, John B., 79
Bouchette, Joseph, 34
Boundary Waters Canoe Area Wilderness, 190–94
Brainerd, 84, 90–91, 104, 179
Brainerd, City of, Crow Wing County, 90
Breckenridge, 111
British and French Dominions in North America, Map of the, 24–25
Brown County, 53, 96–97
Browne County, 56
Burnham, Daniel, 157
Burroughs, H. N., *The United States,* from *Augustus Mitchell, New Universal Atlas,* 45

C

Cabot, John and Sebastian, 34
Campbell, W. P., 84
Cannon River, 46, 78, 112
Carver, Jonathan, 7, 26, 42; *Travels in the Interior Parts of North America,* 26
Carver's Travels in the Interior Parts of North America in 1766 and 1767, A Plan of Captain, 26–27
Cass, Lewis, 29–30, 32, 40
Cass County, 51, 58, 60
Champlain, Samuel de, *Carte de la Nouvelle France . . . ,* 6, 10–11
Chatelain, Henri Abraham, *Carte particuliere du fleuve Saint Louis,* from *Atlas.,* 18–19
Chicago, 64, 128, 146, 157, 178
Chippewa. *See* Ojibwe.
Chisago County, 66, 67; map of, 76, 77
Chisago County, Plat Book of, 6, 62
Clark, William, 29
Cleveland, H. W. S., 136–37, 154–55; *Suggestions for a System of Parks,* p. 154–55
Colton's Minnesota and Dakota, from *Colton's Atlas of the World,* 56–57
Constellation Cities, 1985, from *Report Number 5, April 1968,* 171

Cooke, Jay, 101
Coronelli, Vincenzo, *Partie occidentale du Canada ou de la Nouvelle France,* 16–17
Coteau des Prairies, 46, 61
Cottage Grove, 70, 169
Cottage Grove Township, from *Plat Book of Washington County,* 70
Counties of Minnesota, Map of the Organized, 1850, 46–47
Counties of Pine, Kanabec, Isanti, Chisago, from *Illustrated Historical Atlas,* 76–77
County Map of Minnesota, W.H. Gamble, no.81, from *Mitchell's New General Atlas,* 60–61
Crawford, Andrew W., 131, 157–58, 160
Crocus Hill, 162
Cross Lake, 59
Crow Wing, 54, 61, 90, 111, 112
Crow Wing River, 46, 92

D

Dablon, Claude, *Lac Superieur,* 12–13
Dakota County, from *Illustrated Historical Atlas of Minnesota,* 78
Dakota County, 51, 53, 169
Dakota (Sioux) Indians, 16, 30, 42, 44, 61, 111, 115; and Carver, 26; and cession of lands, 48, 50, 54, 56, 74, 84; and Fort Snelling, 132; and Hennepin, 14–15; and Mille Lacs, 40
Dakota Territory, 56, 58, 111
Dalles of the St. Croix, 66
Dalrymple, Alexander, 142
De Graf, C. A., 74
Detroit Lakes, 73, 111
Devil's Lake, 52, 54
Dodge, Henry, 42
Dodge County, 52, 108
Donaldson, William L., 141
Doty County, 55
Douglas, Senator Steven, 45
Du Luth, Daniel Greysolon, Sieur, 15
Dubuque, 44, 88
Duluth, 60, 73, 100, 101, 102, 104; maps of, 82, 100, 102–3

210

INDEX OF MAPS, PEOPLE, AND PLACES

Duluth, Perspective Map of, 100
Duluth (Minnesota) and Superior (Wisconsin), 102–3
Duluth Street Railway Company, 102–3

E

Echo Trail, 192
Edina, 152, 169, 171, 174, 194
Ellsbury, Geo. H., *St. Paul, Minnesota,* 134–35

F

Falls of St. Anthony. *See* St. Anthony Falls.
Farmer, John, *Improved Map of the Territories of Michigan and Ouisconsin,* 40–41
Farmer, The, 68
Farmers' Atlas & Directory, 1913, 68–69
Fillmore County, 51, 52, 86, 88
Fisher, William A., 192
Fond du Lac, 48, 54, 61, 79, 102
Foote, Charles M., 66
Foote, Ernest B., 67
Forest City, 92
Forest City Merchant Flour Mill, View of, from *Illustrated Historical,* 92
Fort Francis, 51
Fort Gains. *See* Fort Ripley
Fort Ridgely, 60, 113
Fort Ripley, 46, 49, 112; and road network, 54
Fort Snelling, 34, 35, 41, 49, 111; squatters removed from, 132
Fort Snelling, Map of the Military Reserve Embracing, 132–33
Fort William, 54, 111
Fountain Cave, 40, 132
4,000,000 by 2000! Preliminary Proposals for Guiding Change, 171
Freeborn County, 52, 86; map of, 36
Freeborn County, Survey of Township 103, Range 19 West, 1855, 36–37

G

Gilbert, Cass, 70, 160, 162
Glacial Lake Agassiz, 54, 111
Golden Valley Township, 152
Goodhue County, 51, 52, 53

Grand Portage, 51, 54, 61, 111
Grand Rapids, 61
Grant County, 66
Grant County, Wisconsin, Atlas of, 66
Great Lakes, 10, 16, 18, 20, 25
Green, Charles, 150
Gruen, Victor, 171, 174
Guerin, Jules, 157; *The Coming Metropolis: General Perspective Looking Northwest,* 130–31; *The Sixth Avenue Approach to the Institute of Arts,* 158–59; *The Sixth Avenue Artery,* 157
Guide to Studies of Social Conditions of the Twin Cities, 164
Gull Lake, 61
Gunflint Trail, 179, 192

H

Hageboeck, Augustus, 138
Ham Lake, 74
Hamm, Theodore, 144
Harding, Warren, 70
Harriman, J. D., 118
Hastings, 78
Heinselman, Miron, 200
Helm, H. H., 73
Henion, John W., 66
Hennepin, Louis, 14, 20; *Carte de la Nouvelle France et de la Louisiane,* 14–15
Hennepin County, 150, 169
Herrold, George H., 160, 162
Hibbing, 40, 61, 94–95
Hibbing, Frank, 94
Hibbing Quadrangle, Minnesota, 94
High Forest, 53
Hill, James J., 118, 138, 146
Ho Chunk Indians. *See* Winnebago Indians
Hokah River, 46
Homann, Johann Baptist, 20–21; *Amplissimae regionis Mississipi,* 20–21
Hood, Edwin C., 66
Hopkins Plat Book of St. Paul and Suburbs, 146–47
Hopkins, 152, 169
Houghton, George, 32

Hudson Bay, 10, 13, 18, 25, 61, 111
Hutchins, Thomas, 36
Hydrographical Basin of the Upper Mississippi River, 1843, 28, 34–35

I

Illustrated Historical Atlas of Minnesota (Andreas), 72, 73–74, 75, 76–77, 78, 79, 80–81
Indian Country in 1832, Map of the Route Passed over by an Expedition into the, 32–33
Iniskah River, 46
Insurance Maps of Minneapolis (Sanborn), 140–41, 142–43
Iowa and Wisconsin, 1845, 44
Irvine, Larry, 174
Itasca, 86
Itasca County, 51, 58, 60, 184; map of, 76, 77
Itasca County, Minnesota, The Heart of the 10,000 Lakes Country, 184–85
Itasca State Park, 188

J

Jardine, William, 191
Jefferson, President Thomas, 29, 123
Johnson's New Illustrated (Steel Plate) Family Atlas, 38

K

Kanabec County, Map of, 76, 77
Kaposia (Little Crow), 31, 48
Kellogg, Frank B., 70
King, Nicolas, 31
Kirk, T., 64
Knowles, H. B., *Bird's-Eye View of Farm and Residence,* 106

L

La Crosse, 54, 60
Lac Qui Parle, 54
Lake County, 55, 58, 60
Lake Huron, 10, 12,
Lake Itasca, 32, 35
Lake Minnetonka, 149, 150
Lake of the Isles, 152, 156
Lake of the Woods, 16, 23, 25, 26, 56
Lake Okoboji, 181

Lake Pepin, 26
Lake Superior, 10, 12–13, 18, 23, 26, 32, 61, 101, 111
Lake Superior, Jesuit map of, 12–13
Lake Traverse, 50, 54
Lakes Adjacent to the Gunflint Trail, no. 114, of Superior-Quetico Canoe Maps, 192–93
Langdon, 70
Langle, Ed., *Minneapolis, City of Lakes,* 194–95
Langwith's Pictorial Map of Minnesota: Come-to-Minnesota Club, 178, 180–81; *Langwith's Pictorial Minnesota: The Land of 10,000 Lakes,* 176, 177
Larpenteur, Auguste, 112
Le Sueur, 53, 178
Leech Lake, 16, 32, 40, 61, 112
Leetonia, 94
Lewis, Meriwether, 29, 31
Lincoln County, 60
Litchfield, 92–93
Litchfield, Map of, 92–93
Little Crow. *See* Kaposia.
London Magazine, 22, 23
Long, Maj. Steven H., 29, 40
Long Prairie, 49, 50, 54, 112
Long Prairie River, 46
Longfellow, Henry Wadsworth, 155
Lowry, Thomas, 141, 154
Lowry Hill, from *Insurance Maps of Minneapolis,* vol. 3, 140–41
Lynd's Trading Post, 58, 60
Lyon County, 60

M

Mahoning, 94–95
Mankato, 54, 115
Mantorville, 108
Marine Mills, 46
Marschner, Francis J., 198–200; *The Original Vegetation of Minnesota,* 198–99
McCarty, D., 64
McNally, Andrew, 118
Mendota, 45, 50, 78, 84, 112, 132
Menomonie Indians, 49, 51
Mesabi Iron Range, 184

Metro Center '85 Illustrative Site Plan, from *Metro Center '85 Study,* 173
Metro Center '85 plan, 174, 175
Metro Center '85 Study for the Development of Program and Priorities, 173
Michigan and Ouisconsin, Improved Map of the Territories of, 40–41
Michigan, 40, 41, 101
Mille Lacs, 15, 34, 40, 46, 54, 112
Minneapolis, 138–43, 150–59, 165–66, 173–75
Minneapolis, City of Lakes, 194–95
Minneapolis, Downtown, from *Insurance Maps of Minneapolis,* vol. 3, 142
Minneapolis, Minnesota, Showing Park System, 154–55
Minneapolis, Minnesota, View of, 138
Minneapolis, Natural Areas, Middle Segment, 165
Minneapolis, Plan of, 131, 157, 158–59
Minneapolis, Vice Areas, 166
Minneapolis West, 150–51
Minnehaha Park and Parkways, Map of Proposed, 156
Minnesota, from *Johnson's New Illustrated (Steel Plate) Family Atlas,* 58–59
Minnesota, from *The World in Miniature, America.* 1857, 55
Minnesota, Langwith's Pictorial Map of, 180–81
Minnesota, Map of, from *Atlas of Western, Northwestern, and Middle Western States,* 118–19
Minnesota, no. 49, from *Colton's Atlas of the World,* 52–53
Minnesota, Northern, from *Illustrated Historical Atlas of Minnesota,* 72, 79–81, 80–81
Minnesota, The Original Vegetation of, 198–99
Minnesota, Where Dreams Come True, 178–79
Minnesota and Its Resources, 114–15
Minnesota Arrowhead Country, 190
Minnesota Invites You to Live, Work, Play in the Playground of 10,000 Lakes, 178–79

Minnesota Land Use and Cover, 1990s Census of the Land, 200–201
Minnesota River, 35, 46, 51–53, 56, 59, 61, 111–13, 115, 131–32
Minnesota Road Maps, Privately published, 122–25
Minnesota State Highway Department Maps, 126–29
Minnesota State Parks, State Forests, and Recreational Areas, 188–89
Minnesota Territory, 45, 52, 54–55, 111
Minnesota Territory, Map of, from John Wesley Bond, *Minnesota,* 114–115
Minnesota Territory, Map of, no. 36, from *Mitchell's Universal Atlas,* 48, 50–51
Mississippi River, 14, 16, 20, 61, 131, 111–13, 115, 154; maps of, 20–21, 30–31; source of, 21, 25, 29, 30, 31, 32, 184
Mississippi River, Map of the, from Zebulon Montgomery Pike, 30–31
Missouri River, 35, 54, 60, 112
Missouri Territory, 46
Mitchell, mining location, 94–95
Mitchell, Augustus, 45
Mitchell, John, *A Map of the British and French Dominions in North America,* 24–25
Mitchell, map publisher, 39; maps by, 54, 61
Morgan Park, Map of, 102, 104–5
Morgan, J. P., 104, 118
Morse, Jedidiah, 55
Morse, Sidney E., *Iowa and Wisconsin, 1845,* 44
Mower County, 52, 53

N

Native Americans, 10, 40, 54, 56, 59, 111, 114, 116, 131; and land cessions, 39, 42, 48
Natural Areas, Central Segment, St. Paul, 1935, from Calvin Schmid, *Social Saga,* 164
Natural Areas, Middle Segment, Minneapolis, 1935, 165
Nau, Anthony, 31
New France, maps of, 10–11, 14–15, 16–17
New Ulm, 53, 83, 96–99
New Ulm, Brown County, Minnesota, 96–97
Newport, 59

INDEX OF MAPS, PEOPLE, AND PLACES

Newton County, 55
Nicollet County, 51, 53
Nicollet, Joseph Nicolas, 28, 34–35, 42, 44, 46; *Hydrographical Basin of the Upper Mississippi River,* 28–29
North America, A New Map of, from the Latest Discoveries, 22–23
North Shore (Lake Superior), 18, 60, 79, 120
Northwest Angle, 25, 56, 60, 181
Nott house, 141

O

Oberholtzer, Ernest, 191
Ojibwe Indians, 30, 32, 34, 46, 59; and cession of lands, 48, 50, 100, 111; and conflicts with Dakota Indians, 111; villages and trading posts of, 46, 48–49, 61, 74, 79
Olmsted County, 52
Olson, Gov. Floyd B., 178
Ossana, Fred, 150
Oza Windib (Yellow Head), 32

P

Parsons, William E., 160, 162
Partridge, George H., 141
Pembina, 49, 54, 111
Pembina County, 51, 58, 60
Pfaender, Wilhelm and Fred, 96–97
Pierce County, 51, 56
Pig's Eye Lake, 162, 196
Pigeon River, 25
Pike, Lt. Zebulon Montgomery, 29, 131–32, 181; *map,* 30–31
Pine City, 59, 76
Pine County, 53, 56, 59, 76; *map,* 76, 77
Plympton, Major Joseph, 132
Point Douglas, 46, 115
Polk County, 58
Pool, 94–95
Pope, Lt. John, 29, 46
Port Charlotte, 51, 54
Prairie du Chien, 45, 111
Prairie Island, 84
Prairie la Belleview, 48, 50
Pratt, Charles, 136
Predominant Land Use 1958, 168
Preliminary City Plan, from City Planning Board, 160–61
Preston, Fillmore County, Minnesota from *Illustrated Historical Atlas,* 86–87
Prignitz, Christian, map by, 96–97

R

Rainy Lake, 26, 51, 54, 191
Rainy River fur-trade route, 111
Ramsey, Governor Alexander, 108, 114
Ramsey County, 53, 56, 169
Rand, William H., 118
Red Ball Route, 123
Red Lake, 23, 61; bog region around, 79; Ojibwe Indian agency at, 48
Red River of the North, 50, 51, 54, 61, 111
Red River Trail, 46, 49, 54, 79, 111, 114
Red River Valley, 58, 60, 79, 113, 114, 138, 165
Red Wing, Goodhue County, Minnesota, from *Illustrated Historical Atlas,* 85
Redwood County, 58, 60
Rice, Henry M., 112
Rice County, 52
Rice Point, 100, 102
Rifkin, 148
Riverfront Housing and Cultural Center, 174–75
Roads in the Territory of Minnesota, Map of the General Government, 112–13
Rochester, 54, 56, 88, 109, 113, 120, 181
Rochester, Olmsted County, Minnesota, from *Illustrated Historical Atlas,* 88–89
Rock County, 56
Roe, Frederick B., *Albertson's Map of the City of Duluth, St. Louis County,* 101
Roosevelt, President Theodore, 70, 118
Root River, 46, 86
Rosemount, 78
Route Passed over by an Expedition into Indian Country, Map of the, 32, 34
Rum River, 26, 34, 112

S

Sanborn, *Illustrated Historical Atlas of Minnesota,* 140–41, 142, 143
Sand Lake, 48
Sandy Lake, 54
Sauk Rapids, 46
Sault Ste. Marie, 101
Sawtooth Mountains, 46
Saxton, 60
Schapekahm, H. G., *Map of New Ulm, Brown County, Minnesota,* 98–99
Schell, August and Otto, 96
Schmid, Calvin F., 164
Schoolcraft, Henry, 29, 32, 34, 40
Schoolcraft Island, 34
Scott County, 53
Scranton, 94
Section 32, Township 29, Range 22, plate 14, from *Atlas of the City of St. Paul,* 144–45
Selkirk Colony, 132
Severance, Cordenio A., and Mary A. Harriman, 70
Shafer Township, from *Plat Book of Chisago County, Minnesota,* 66–67
Shafer, 66–67
Shakopee, 48, 50
Shultz, George, 80
Sibley, Henry, 132
Sibley County, 51, 53
Sioux Treaty of 1851, 48
Sioux. *See* Dakota.
Sisseton, 50, 54
Sisseton Indian lands, 58
Smith, A. C., 64
Smith, John Gregory, and Ann Eliza Brainerd, 90
Smith, Lt. E. K., 132
Smith, William R., *Observations on the Wisconsin Territory,* 42
Snake River 59, 76
Social Saga of Two Cities: An Ecological and Statistical Study of Social Trends, 164
South St. Paul, 146–47, 169, 178
Spirit Lake, 34, 104
St. Anthony, 59, 115, 158, 194
St. Anthony Falls, 14, 16, 26, 30, 44, 45, 132, 136, 138, 149, 175
St. Anthony Park, a Suburban Addition to St. Paul and Minneapolis, Plan of, 136–37
St. Charles, 59, 106, 134
St. Croix Falls, 46
St. Croix River, 30, 131; valley, 45, 46, 49, 66, 112

St. Francis (Rum) River, 15, 26
St. Louis, 16, 30, 102
St. Louis Bay, 102
St. Louis County, 55, 58, 60, 101
St. Louis Park, 152, 169
St. Louis River, 23, 54, 100, 112, 115; *map,* 18–19
St. Paul, 26, 46, 49, 59, 74, 84, 88, 111, 114, 115, 123, 132–37, 144–48, 160–64, 166–67, 196–97
St. Paul, Atlas of the City of, 144–45
St. Paul, Bird's-Eye View of, 74
St. Paul, Capital City of Minnesota, Plan of, 160–61
St. Paul, City Plan, 162–63
St. Paul, Minnesota, by Ellsbury, 134–35
St. Paul, Natural Areas, Central Segment, 164–65
St. Paul, Predominant Land Use, 168
Saint Paul, Souvenir Guide Map of, 166–67; illustrated map, 196–97
St. Paul, Township 28 North, Range 22 West, from *Hopkins Plat Book,* 146–47
St. Paul Zoning Map, 1922, from *City Planning Board . . . ,* 160–61
St. Peter, 88, 109
St. Peter's River, 26, 30
State Highway Department Map of Minnesota, 126–27
Steele County, 49, 52
Stevens, Governor Isaac, 30, 54; *Preliminary Sketch of Northern Pacific Rail Road,* 116–17
Stickney, Alphas P., 146
Stillwater, 46, 54, 111
Summerfield, 70, 71
Superior, Wisconsin, 83, 100, 102
Superior County, 55
Superior National Forest, 179, 190; *map,* 191
Superior-Quetico Canoe Maps, 192–93
Survey of Township 103, Range 19 West, Freeborn County, 1855, 36–37
Svenska Dalen, 144
Swede Hollow, map, 144–45
Swede Mill, 76
Swede Prairie Township, from *Farmers' Atlas & Directory,* 1913, 68–69

Taylor, President Zachary, 45
Taylors Falls, 66
Temperance River, 61
Thompson, Lt. J. L., 132
Thompson, Mr., 31
Toombs County, 56, 58
Tower, Charlemagne, 80
Tracy, Alexandre de Prouville, marquis de, 12
Traverse des Sioux, 48, 50, 83, 112, 115
Twin Cities, 134, 148–150, 156, 165, 168–72
Twin Cities, 2000 AD: Alternative Patterns, from *Report Number 4,* 172
Twin Cities, Metropolitan Council, 131, 168, 169, 172
Twin Cities, Their Famous Lakes, River, Parks and Resorts, 148–49
Twin Cities Area Metropolitan Development Guide, 172
Twin Cities Metropolitan Area Generalized Land Use 1958, 168; *1968,* 169
Twin Cities Metropolitan Planning Commission, 168, 171
Two Harbors, 83, 120

U

U.S. Army, 29, 41, 66, 102, 111, 112, 131
U.S. Congress, 29–30, 36, 56, 102, 111, 191, 192
U.S. Geological Survey, map by, 150
U.S. War Department, 29, 116, 132
United States, Map of the, 45
Upper Mississippi River, 35
Upper Red Cedar (Cass) Lake, 32

V

Vermillion River, 46, 112
Vermilion, 79–80
Vero, *S. St. Paul,* 196–97
Vice Areas, Minneapolis, 1936, 166
Voyageurs Highway, 111

Wabasha, 112
Wabasha County, 51, 52
Waldseemüller, Martin, 7; globe gores, 8–9
Walker, Martin O., 88
Walker, Thomas B., 141
Warner, George, 66

Waseca, 49
Washington County, 70, 169
Washington County, Plat Book of, 70
Washington, D.C., 199; as prime meridian, 40
Wasioja Township, Map of, from *The County of Dodge, Minnesota,* 108–9
Watonwan County, 56
Watonwan River, 53
Webb, Edward A., 68
Wellge, H., *Perspective Map of Duluth, Minnesota,* 100, 102
Wesleyan Methodist Seminary, Wasioja, Dodge County from *Historical Atlas,* 108
West St. Paul, 78, 169
White Bear Lake, 26
Whitewater, 64
Whitewater Falls, 107
Whitewater River, 46; erosion, 106–7; state park, 107; valley, 65, 106–7
Whitewater Township, from *Map of Winona County,* 64
Wilkin, Col. Alexander, 58
Wilkin County, 58
Wilkinson, Gen. James, 30
Winnebago Indians (Ho Chunk), 46, 49, 50, 112
Winona, 54, 59, 60, 64, 88, 106; bird's-eye view of, 74
Winona County, 64–65
Winston City, 79–80
Wirth, Theodore, park superintendent, 154
Wisconsin, 18, 41, 42, 45, 120, 125
Wisconsin Territory, Map of the Settled Part of, 1838, 42–43
Wonderland Trail, 123
Woods Trail, 111

Y

Yellow Medicine County, 68
Young, J. H., *Map of Minnesota Territory, no. 36,* from *Mitchell's Universal Atlas,* 48–49, 50–51

Z

Zenith City, 101
Zumbro River, 46, 108

✴ CREDITS ✴

The library of the Minnesota Historical Society, St. Paul, holds the maps and other images in this book with these exceptions, which are reproduced with permission: pages 45, 48–49, 50–51, 52–53, 55, 56–57, 59, 61, 95, 98–99, 113, 114-15, 128–29, 130, 144–45, 151, 157, 158–59, 160–61, 162–63, 164, 165, 166, 168, 170, 171, 172, 173, 174–75, 176, 181, 186–87, 190, 191, 193, 195, 196–97, 201, David Lanegran Collection; pages 8–9, 11, 13, 24, 25, James Ford Bell Library, University of Minnesota, Minneapolis; pages 96–97, Brown County Recorder's Office, New Ulm; pages 103, 104-105, Northeast Minnesota Historical Center, Duluth; page 133, Hennepin County Surveyor's Office, Minneapolis; pages 122–23, 124–25, 126, 215, Craig Solomonson Collection; pages 136–37, 167, Ramsey County Historical Society, St. Paul; pages 146–47, Dakota County Historical Society, South St Paul; pages 154–55, 156, Minneapolis Public Library; pages 178, 179, 184, 185, 186, 209, Laura Kigin Collection; page 187, Lured to the Lake, Minnetonka; page 204, "Minnesota," from *Encyclopedia Britannica* (New York: Gil Edgewood Co., 1910).

Minnesota on the Map
was designed and set in type
at Cathy Spengler Design, Minneapolis.
The typefaces are Malaga and Honduras.
Printed by C & C Offset Printing.